INTERFERENCE CONTROL
IN CABLE AND DEVICE INTERFACES

by

CHRIS J. GEORGOPOULOS

Interference Control Technologies, Inc.
Route 625, Gainesville, VA 22065

© 1988 by Interference Control Technologies, Inc.
All rights reserved. Published 1988
Printed in the United States of America
95 94 93 92 91 90   5 4 3 2

Library of Congress Catalog Card Number: 87-81843
ISBN: 0932263-26-7

This book, or any part thereof, may not be reproduced
in any form without written permission from the publisher.

Acknowledgement

The preparation of this book has left me indebted to many people for their advice, assistance and suggestions. First, I wish to thank Mr. Donald R. J. White, president of Interference Control Technologies (ICT) for providing the enthusiastic encouragement, friendly criticism and discussion that every book author needs and appreciates. I also wish to thank all personnel of the Publications Department of ICT, not only for the typing of the handwritten manuscript, but also for their successful pursuit of the innumerable details involved in organizing the art work, etc.

I would like to express my sincere appreciation to several individuals of the Electrical Engineering Department of the University of Thrace, Greece, and especially to Mrs. H. Kondili and Messrs. B. Bakirtzis and G. Vaidis for their help. Also, a number of students at the University of Thrace, Greece, to whom part of this book was taught, deserve recognition and thanks for their cooperation.

Besides the credit that is given in each case throughout this book for any material used, many thanks are herein extended to the companies, editors of various publications and the authors of many excellent papers on electromagnetic interference and related topics who have granted permission for using such material.

I want to acknowledge the understanding and consideration given me by my wife and my two daughters while writing this manuscript. Finally, I dedicate this book to my parents who have instilled in me the necessary confidence to dream, investigate and implement ideas in real life.

Prof. Chris J. Georgopoulos
Brief Biography

Dr. Chris J. Georgopoulos has been educated in both Greece and the United States and possesses the following degrees: BS, BS(EE), MS(EE) and Ph.D(EE).

From 1963 to 1966 he was employed by Sylvania Electric Products in Danvers, MA, as Design and Development Engineer, and from 1966 to 1967 by Wang Laboratories in Tewksbury, MA, as Senior Design and Applications Engineer.

In 1967 he joined the Raytheon Digital Systems Laboratory, Bedford, MA. He held the positions of Group Leader and Section Manager and was responsible for the design and development of interface circuits and testing equipments for phased array radar systems. In 1972 he joined the Technical Research Staff of the University of Patras where he worked on a number of research projects until 1976.

In 1976 he became Professor and Head of the Electrical Department of the School of Engineering Technologists at the K.A.T.E. Technical Center of Patras, Greece. In 1977 Dr. Georgopoulos was elected full professor of the Chair of Electronics and Director of the Electronics Laboratory in the School of Engineering of the University of Thrace, Xanthi, Greece. From March 1978 to August 1979 he was the Dean of the Engineering School of the University of Thrace. Presently he is the Electronics Division Chairman and Director of the Electronics and Digital Systems Laboratories.

Professor Georgopoulos has received five patents covering his inventions and several awards for his innovative and cost reduction ideas; has written some 40 technical reports on research,

Biography

design and development projects; has published and presented more than 50 technical papers and has authored six books in the field of electronics and digital systems (in the Greek language). He is also the author of two other books of worldwide circulation: *Fiber Optics and Optical Isolators* (DWCI: VA, USA, 1982), and *Interface Fundamentals in Microprocessor-Controlled Systems* (D. Reidel Publishing Co.: Dordrecht, Holland, 1985). He is also a member of the editorial board of Electrosoft, an international quarterly journal of the Computational Mechanics Publications, Southampton, England.

His research and development interests include fiber optics and IR communications, local area networks, special computer interfaces and automated office hardware. He has been a consultant in the above areas in both Europe and the USA, and during the Academic Year 1986–1987 was visiting professor in the Electrical Engineering Department of the University of Lowell, Massachusetts.

Dr. Georgopoulos is a Professional Engineer in the State of Massachusetts, a member of the Technical Chamber of Greece, a senior member of the IEEE, a member of the International Democritos Institute, a member of the IEEE Optical Communications Committee and a member of the E.A.M.E.C. (European, African and Middle East Committee) of the IEEE Communications Society.

Preface

As electronic devices and systems become more sophisticated and more numerous, the airspace becomes increasingly polluted with *electromagnetic radiation*. On one hand, electrical products that are capable of behaving like antennas generate electromagnetic emissions which interfere with the operation of electronic equipment. On the other hand, this equipment, especially digital equipment, produces high-frequency signals which hamper the performance of other sensitive electronic equipment and devices.

Interface is a shared boundary between elements defined by common physical interconnection characteristics, signal characteristics and meanings of interchanged signals subjected to interference problems as are other elements in the system which it interconnects. The protection of equipment, its interface(s) and the environment from *electromagnetic interference (EMI)* is a necessity in the commercial, industrial, aerospace and military fields.

In electronics, the meaning interference protection is covered by the field of *electromagnetic compatibility (EMC)*, i.e., the ability of electronic (or electrical) devices or systems to operate in their intended electromagnetic environment at designed levels of efficiency. At present, references to the various facets of EMC performance of devices, equipment and systems are dispersed throughout the literature, and designers in the field must obtain a large part of their information through manufacturers' component data sheets and application notes.

In this book a coherent approach is developed by applying the information available in the numerous publications to solve potential interface EMI problems. Written with a minimum of

Preface

mathematics and illustrated with tables, suitable graphs and figures, this is an easy-to-understand text. It gives practicing engineers, engineering students and technicians a full understanding of the problems and detailed explanations of the principles and design approaches necessary to control interference in cables and device interfaces.

The first chapter is an introduction to the interface domain and EMC problems facing the engineering community. Chapter 2 deals with the basic logic families in terms of their performance, capabilities and behavior under interference conditions. Line drivers/receivers, applicable standards and interfacing cables are discussed in Chapter 3 along with basic transmission line principles. Chapter 4 focuses on terminal and peripheral interfaces, while Chapter 5 deals with various functional aspects of bus interfaces, the connection of devices to a bus line and solutions to noise problems.

Chapter 6 discusses the basic building blocks that constitute a data acquisition system and the problems associated with the various interfaces and interconnecting links when operating in an industrial environment. Chapter 7 shows that effective noise detection and measurements involve careful selection of instrumentation, proper design of test fixture interfaces and knowledge of EMI test limits for compliance to holding standards. In Chapter 8, various interfaces of optoelectronic systems are discussed, including: optical isolators, fiber optic links and infrared transmission equipment in confined spaces. As in all electronic installations, unwanted electromagnetic energy can be conducted and radiated into computing and other electronic equipment of a local area network (LAN) system, causing excitation of sufficient amplitude to disturb its operation; techniques for reducing such effects are discussed in Chapter 9. Finally, Chapter 10 covers RF and microwave equipment interference problems and provides some interference control guidelines.

Xanthi, Greece
February, 1988

Dr. Chris J. Georgopoulos
Professor

Table of Contents
Interference Control in Cable and Device Interfaces

	Page
Acknowledgement	iii
Other Books Published by ICT	iv
Preface	xv
Illustrations and Tables	v
Abbreviations and Symbols	xxxi

Chapter 1 An Introduction to the Interface Domain and EMC Problems 1
- 1.1 Introduction 1
- 1.2 Interface Basics 2
    - 1.2.1 Some Useful Definitions 2
    - 1.2.2 Understanding and Using Interface Specifications 3
    - 1.2.3 Interface Circuit Selection for a Digital System 4
    - 1.2.4 The Importance of Interfacing and the EMI Problem 6
- 1.3 Review of Interference Problems 8
    - 1.3.1 Definitions of EMC-Related Terms 8
    - 1.3.2 Coupling Between Interference Source and Receptor 10
    - 1.3.3 EMC Performance of Electronic Cables 11
    - 1.3.4 The ESD Problem 13
    - 1.3.5 Challenges Facing the EMC Community 13

1.4 Standard Limits and Measurements.................. 15
    1.4.1 FCC Sets the United States Guidelines....... 15
    1.4.2 VDE Sets the European Guidelines.......... 16
    1.4.3 CISPR's Recommendations.................. 18
    1.4.4 Interference Measurement Methodology...... 19
1.5 References.. 22

Chapter 2 Logic Elements Interfacing............... 25
2.1 Introduction...................................... 25
2.2 Review of Basic Logic Families 26
    2.2.1 TTL Family................................ 26
    2.2.2 ECL Basics................................ 31
    2.2.3 The CMOS Family......................... 33
    2.2.4 GaAs ICs.................................. 37
    2.2.5 Technology Comparison 39
2.3 Interconnection Approaches and Minimization of Interference Problems 42
    2.3.1 Bandwidth Demands of Fast Logic 42
    2.3.2 TTL ICs in Industrial Environments 44
    2.3.3 Wiring ECL Gates and GaAs Devices at the Printed Circuit Board Level................. 47
    2.3.4 CMOS Interconnections.................... 51
    2.3.5 EMI Problems in Microelectronics of High Complexity................................ 54
    2.3.6 Susceptibility of ICs to Electrostatic Discharge Damage 55
2.4 Interfacing Different Logic Families—Design Examples... 58
    2.4.1 Brief Compatibility Comparisons............ 58
    2.4.2 ECL/TTL Interfaces........................ 60
    2.4.3 CMOS/TTL Interfaces..................... 62
    2.4.4 CMOS/ECL Interfaces..................... 64
    2.4.5 HSCMOS/CMOS Interfaces 65
2.5 References.. 67

Chapter 3 Line Drivers/Line Receivers and Interfacing Cables 71
3.1 Introduction...................................... 71
3.2 Signal Transmission Lines......................... 72
    3.2.1 Definitions of Line Transmission Factors..... 72
    3.2.2 Digital Signal Line Models................... 73
    3.2.3 Line Termination with Resistive Element..... 76
    3.2.4 Line Termination with Clamping Diodes...... 78

Contents

- 3.3 Problems with Interfacing Cables and Connectors... 79
  - 3.3.1 Basic Types of Cable Lines and Their Characteristics... 80
  - 3.3.2 Figure of Merit for Shielded Twisted Pair and Coaxial Cables... 85
  - 3.3.3 Ground Loops... 86
  - 3.3.4 Crosstalk and Common-Mode Rejection... 88
  - 3.3.5 EMI/RFI Control at the Connector Interface... 89
- 3.4 Characteristics and Applications of Line Drivers/Receivers... 90
  - 3.4.1 Description and Operation of Typical Line Drivers and Line Receivers... 91
  - 3.4.2 EIA and Other Standards for Line Drivers/Receivers... 93
  - 3.4.3 Transceivers and Repeaters... 97
- 3.5 References... 100

Chapter 4 Terminal and Peripheral Interfaces... 103
- 4.1 Introduction... 103
- 4.2 Terminal Interfaces: Some Definitions and Standards... 104
  - 4.2.1 Some Basic Definitions... 104
  - 4.2.2 Common Data Interfaces for Digital Communications Links... 105
  - 4.2.3 International Modem Standards... 108
  - 4.2.4 The EIA RS-449A Data Communications Interface... 111
- 4.3 Terminal Interfaces: Representative Design Examples... 112
  - 4.3.1 Man/Machine Interface... 113
  - 4.3.2 DCE-to-DTE Interconnections with Balanced/Unbalanced Circuits... 115
  - 4.3.3 CPU-to-Remote CRT Terminal Interface... 117
- 4.4 Peripheral Interfaces... 119
  - 4.4.1 Some Useful Definitions... 119
  - 4.4.2 Peripheral Interface Chips... 119
  - 4.4.3 Disk and Tape Drive Interfaces... 122
  - 4.4.4 Interface for Micro-to-Mini Data Transfers... 124
  - 4.4.5 Peripheral Interface Controllers... 125
- 4.5 Interface Interference Problems and Solutions... 128
  - 4.5.1 Noise Suppression in Computer Grade Connectors and Peripheral Actuators... 128

	4.5.2	Conductive Shield Termination at a Terminal or Peripheral Device.......................... 130

       4.5.2 Conductive Shield Termination at a Terminal or Peripheral Device.......................... 130
       4.5.3 Use of Optocouplers as Interface Isolators ... 131
       4.5.4 EMP Radiation Detector Circuit for Terminal and Peripheral Protection 132
4.6 References 135

Chapter 5 Data Bus Interfaces 139

5.1 Introduction 139
5.2 The STD Bus 140
       5.2.1 Bus Description........................... 140
       5.2.2 Compatibility 141
       5.2.3 Real-Time Applications 142
       5.2.4 Intelligent Interfaces for STD Bus Systems ... 142
5.3 General Purpose Interface Bus (GPIB) or IEEE 488-1978 143
       5.3.1 Definitions and Equivalent GPIB Standards... 143
       5.3.2 Incompatibilities and Limitations............ 146
       5.3.3 Using Expanders to Extend the IEEE-STD-488 Bus Capabilities 147
5.4 Multibus System Bus (IEEE-P796)................ 148
       5.4.1 Definitions and Evolution of Multibus Systems................................. 148
       5.4.2 Multibus II Products....................... 149
       5.4.3 Signal Conditioning of Multibus Compatible Analog I/O Modules 150
5.5 The VME Bus................................... 152
       5.5.1 VME System Features 153
       5.5.2 VME System Architecture.................. 154
       5.5.3 VME Bus and Multibus II Comparison 155
       5.5.4 PC/VME Master Interface System 156
5.6 Other Types of Data Buses 158
       5.6.1 Brief Descriptions and Special Features...... 158
       5.6.2 Some Comparisons........................ 159
5.7 Device Connections and Interference Problems in Bus Organized Systems........................ 163
       5.7.1 Bus Transformer Coupling in MIL-STD-1553B 163
       5.7.2 TTL-System Bus Line with Combined Active/Passive Terminations 164
       5.7.3 RFI and EMI Reduction with Differential Backplane Transceivers.................... 166
5.8 References 170

Contents

Chapter 6 Data Acquisition Systems Interfaces 173
6.1 Introduction 173
6.2 Basic Building Blocks of a Data Acquisition System 174
    6.2.1 Data Acquisition System Architecture 174
    6.2.2 Basic Interface Design Problems 175
    6.2.3 Input Multiplexer 176
    6.2.4 Programmable Gain Amplifier 177
    6.2.5 Sample and Hold Amplifier 177
    6.2.6 Analog-To-Digital Converters 179
    6.2.7 Digital-To-Analog Converters 181
    6.2.8 Timing Considerations 183
    6.2.9 DAS-to-Microprocessor Input/Output Interface 183
6.3 Data Acquisition and Data Interface Considerations 184
    6.3.1 A Microcomputer-Operated Process Control System 184
    6.3.2 Source Characteristics 187
    6.3.3 Data Acquisition From Thermocouple Inputs 188
    6.3.4 Voltage-To-Frequency and Frequency-To-Voltage Converter Links 190
    6.3.5 Single-Line Power and Data Transmission Link 192
6.4 Protection of Data Acquisition Systems from Interference 193
    6.4.1 Alternative Techniques for Voltage Isolation and Analog Data Transmission from Remote Transducers 193
    6.4.2 Input Amplifiers and Multiplexer Section Conditioning 195
    6.4.3 Problems With A/D and D/A Subsystems 198
    6.4.4 Data Relaying Section 201
    6.4.5 Hazardous Areas and Temperature Effects ... 202
6.5 References 204

Chapter 7 Measurement and Automation Interfaces .. 207
7.1 Introduction 207
7.2 Instrumentation Amplifiers and Input Noise Problems 208
    7.2.1 Definitions and General Characteristics 208
    7.2.2 Common-Mode Rejection 211

		7.2.3	Isolation Amplifiers for Effective Measurements	213
	7.3	Testability and Test Fixtures		215
		7.3.1	Built-in Testability	215
		7.3.2	Test Fixtures and Interference Problems	216
		7.3.3	Guidelines for Fixture Design and Interconnections	220
	7.4	Analyzers and Automatic Test Interfaces		221
		7.4.1	Dedicated Logic Analyzers	221
		7.4.2	Receivers and Spectrum Analyzers for RFI/EMI Measurements	222
		7.4.3	GPIB-Based Automated Systems	224
	7.5	FCC Rules and Digital Equipment: Testing for Compliance		226
		7.5.1	FCC Computing Device Rules	226
		7.5.2	The FCC Standard of Measurement of Radio Noise Emissions from Computing Devices	227
		7.5.3	Conducted-Emission Tests	228
		7.5.4	Radiated-Emission Tests	230
	7.6	References		232
Chapter 8		**Optoelectronic Systems Interfaces**		**235**
	8.1	Introduction		235
	8.2	Optical Isolators		236
		8.2.1	Basic Types and Applications	236
		8.2.2	Digital Links With Optical Isolators	237
		8.2.3	Linear Operation	238
		8.2.4	Example of Laser Power Supply Isolation	240
	8.3	Cables and Connectors in Fiber Optic Link		241
		8.3.1	Typical Fiber Optic Link	241
		8.3.2	Fiber Optic Cable Advantages	242
		8.3.3	Types and Transmission Properties	242
		8.3.4	Environmental Criteria for Fiber Optic Cables	244
		8.3.5	Fiber Optic Sensors	245
		8.3.6	Fiber Optic Connectors Control EMI/RFI	247
		8.3.7	Radiation Effects on Fiber Optics	249
	8.4	Light Sources and Detectors in Fiber Optic Links		250
		8.4.1	Spectral Matching of Fiber Optics to Sources and Detectors	250
		8.4.2	Light Sources and Their Interfaces	251
		8.4.3	Photodetectors and Their Interfaces	253
		8.4.4	Snap-In Fiber Optic Links and EIA Interface Compatibility	255

Contents

      8.4.5   Flux and Bandwidth Budgeting 257
  8.5 Wireless Links Via Diffuse Infrared Radiation 258
      8.5.1   Infrared Link for Automated Factory
            Environment 258
      8.5.2   Sources of Interference and Other Detection
            Problems................................... 259
      8.5.3   Background and Transient Light Effects...... 260
  8.6 References... 261

Chapter 9 Local Area Networks: Interfaces and Interconnections 265

  9.1 Introduction....................................... 265
  9.2 Definitions and Transmission Media for LANs........ 266
      9.2.1   Some Fundamental Concepts 266
      9.2.2   Implementation of Alternative Transmission
            Media..................................... 273
      9.2.3   Comparison of Alternative LAN Transmission
            Media..................................... 277
      9.2.4   Standardization: The ISO Open
            Interconnection Model..................... 280
  9.3 Controlling Interference in LAN Systems 281
      9.3.1   Data Processing Industry LANs: The
            "Antenna Farm".......................... 282
      9.3.2   Power Supply Line-to-Load Conditioning 284
      9.3.3   Interface Devices and CPU-Terminal/
            Peripheral Interconnection Lines 285
      9.3.4   Noise Immunity Improvement by Correct
            Choice of Protocols 285
      9.3.5   Installed System Cables vs. Environment..... 289
  9.4 Monitoring and Alarm Systems Interfaces 290
      9.4.1   Automatic Monitoring and Restoration....... 290
      9.4.2   Alarm Management in Distributed Control
            Systems................................... 292
      9.4.3   Fault Bypassing in Ring Configuration with
            Alternate Ring 292
  9.5 References... 295

Chapter 10 RF and Microwave Systems Interfaces... 299

  10.1 Introduction...................................... 299
  10.2 The Friendly Field of RF and Microwaves.......... 300
      10.2.1 The ANSI Safety Standard for Exposure to
            RF and Microwaves....................... 300

  10.2.2 New UK Limits for Exposure to RF and
    Microwaves 302
  10.2.3 Therapeutic Effects and Shielding
    Requirements........................... 303
  10.2.4 RF/Microwave Applicators 304
10.3 Impedance Matching and Balun Interfaces 304
  10.3.1 Effects of Mismatch on Amplifier
    Performance at High Frequencies............ 305
  10.3.2 Simple Matching Networks................. 305
  10.3.3 Balun Transformers 307
10.4 RF/Microwave Communication Equipment
  Interference Problems 312
  10.4.1 The General Interference Problem........... 312
  10.4.2 Near-Field Communication 313
  10.4.3 Radiating RF Cables...................... 314
  10.4.4 Microwave Transmission System for LANs ... 315
  10.4.5 Aerospace EMC: Tests Specifications
    Comparison 317
  10.4.6 Some Guidelines for Avoiding Interference
    Problems in Microwave Designs 320
10.5 References..................................... 322
Index 325

Biography 335

Illustrations

Figure	Title	Page
1.1	Processor-to-I/O Interface	2
1.2	General Illustration of Connection of Data Transmission in a Communications-Served Data Processing System	3
1.3	Interface-Circuit-Selection Decision Tree	6
1.4	EMR Incident Upon System is Coupled Through Apertures to System Interior	11
1.5	Radiation Can Originate from Several Places on a Wire Object. Here They Are Depicted Conceptually and Include: (a) The Source Region, (b) An Impedance Load, (c) A Change in Radius, (d) Sharp Bend (or a Junction), (e) A Smooth Curve, and (f) An Open End	12
1.6	Challenges Facing the EMC Community	14
2.1	Typical TTL NAND Gate	26
2.2	Voltage Characteristics Required at the Output and Input of Typical TTL Family Logic	27
2.3	Tri-State Gate Structure	28
2.4	Schottky TTL Logic Circuit	29
2.5	Basic ECL Gate Structure	31

Illustrations

2.6	Specification Points for Determining Noise Margin in ECL Logic Family	33
2.7	Two-Stage Implementation of the Exclusive—OR Function	34
2.8	Typical CMOS Input and Output Characteristics at a Power-Supply Voltage of 5 V	34
2.9	Equivalent Circuits of GaAs Logic Gates	38
2.10	Dissipation for Gate vs. Gate Switching Speed for Various Logic Families	39
2.11	Relative Noise Immunity Comparison for 74LS and 74HC System Design	40
2.12	Bandwidth, Pulse Rise Time and Critical Wavelength Plotted Relationships	43, 44
2.13	(a) Stray Circuit Elements and (b) Common Impedances in Connections of TTL Circuits	45
2.14	(a) Emitted and (b) Conducted Interference Levels for TTL Gate Compared to FCC Specification Limits	46
2.15	ECL-to-ECL Connections	48
2.16	Various Types of Transmission Lines for Printed Circuit Board Connections	50
2.17	(a) Circuit Diagram of CMOS Inverter and (b) Voltage Transfer Characteristic of CMOS Inverter	52
2.18	Indentation of Power-Supply Connections Ensures Proper Sequence of PIN Engagement	53
2.19	SCR Protection Structure Incorporating Transient-Current Sense Resistor	56
2.20	CMOS ESD Protection Circuits: (a) Standard-Gate-Oxide Protection Network, (b) Improved Gate-Oxide Protection Network and (c) Protection Circuit Permitting Level-Shifting Function	57
2.21	Compatibility Requirements of Various Logic Families	60

2.22	Common Supply ECL/TTL Interface Circuits	61
2.23	CMOS/TTL Interfaces	63
2.24	(a) CMOS-to-ECL Interface and (b) Level Translator	64
2.25	(a) ECL-to-CMOS Interface and (b) Level Translator	64
2.26	(a) Standard CMOS-HSCMOS Interface with Up/Down Voltage Conversion and (b) Standard CMOS-HSCMOS Interface With High-to-Low Voltage Conversion	66
3.1	A Digital Signal Line With Coaxial Cable	74
3.2	A Digital Signal Line With Twisted Pair Line	75
3.3	Signal Edge Rise Time vs. Distance in Two Interconnection Media	76
3.4	(a) Impedance Termination in Transmission Line, (b) The Parallel Combination of Termination Resistor (R_T) and the Input Resistance (R_{in}) of the Line Receiver Makes Resistor R_L	77
3.5	Common Diode Terminated Interconnection	78
3.6	Schottky Diode Termination of a Line	79
3.7	Ribbon Cable Applications: (a) Alternating Signal and Ground Return Lines, (b) Differential Drive	81
3.8	Grounded Coax Cables	83
3.9	Shielded Cables for High Interference Environments	84
3.10	Ground Loops Formed by Interconnecting Electronic Equipment in Different Buildings	86
3.11	Elimination of Ground Loops	87
3.12	(a) Differential Amplifiers Reduce Crosstalk by Transmitting the Useful Signal While Cancelling Noise Common to Both Inputs and (b) Amplifiers Are Used at Both Ends of a Transmission Line in a Driver/Receiver Configuration	88

Illustrations

3.13	Connector Feedthrough Capacitor. Effective Bypassing of RF Currents (a) Depends Mainly on Capacitor, But (b) Parasitic Inductor in Equivalent Circuit Limits Bypassing at High Frequencies	90
3.14	Typical Schematic of a Line Driver	91
3.15	Simplified Schematic of a Typical Line Receiver	92
3.16	Circuit Connections and Devices for EIA Standards	96, 97
3.17	Dual Differential Driver/Receiver (Repeater Model 396)	99
4.1	Typical Data-Transmission System	104
4.2	Three Alternatives for Data Transmission Between a Private Branch Exchange (PBX) and a Host Computer	110
4.3	Relative Cost Per Number of Channels Shown in Fig. 4.2	111
4.4	DCE/DTE Interconnection Diagram Illustrating Ground Arrangements for Balanced and Unbalanced RS-449 Interchange Circuits	116
4.5	CPU-to-Remote CRT Terminal Interface and Other Devices With Bus-Line Bidirectional Repeater	118
4.6	Disk Drive Interfaces Span Wide Performance Range	122
4.7	Streamer Buffered Interface (SBI) Board	124
4.8	Block Diagram of Hard Disk Controller	126
4.9	Transient Suppression by ITT-Cannon Connector Having a Diode Chip on Each Contact	128
4.10	Clamping Diodes at Mounting Connectors of Ribbon Cable at I/O Devices	129

4.11	Noise Suppression Techniques in Peripheral Actuators With: (a) Common Diode, (b) Zener Diodes, (c) Varistor, (d) Shunt Current Around Inductance, (e) Circuit Extending Dropout Time	130
4.12	Conductive Shield Termination Encloses Conductors and Extends Enclosure Shield to Cable Shield	131
4.13	Optical Interfaces: (a) Low-Speed Optocoupler and (b) High-Speed Optocoupler	132
4.14	Hybrid Nuclear Event Detector Chip; HSN-3000	134
5.1	STD Bus System	140
5.2	GPIB/IEEE-488 Active Signal Lines for Multiple Devices	144
5.3	Expanding GPIB's Load Capabilities via Bus Amplifier	147
5.4	Block Diagram of Multibus System Bus With External Signal Conditioning	151
5.5	Multibus Board-Level Products Require Interfacing to the Industrial Environment. Analog Devices' RTI-711 Provides 16 Single-Ended or 8 Differential Analog Inputs With 12-Bit Resolution	152
5.6	VME System Architecture is Configured Around Three Bus Structures; The VME Bus, and the VMS Bus	155
5.7	Board Combo Matches PC-to-VME Bus	157
5.8	Data Bus Interface Using (a) Transformer Coupling and (b) Direct Coupling	164
5.9	(a) Normal and (b) Inverted Transistor Operating Modes	165
5.10	Bus-Line System With Combined Passive/Active Terminations	166
5.11	Termination of Am 26LS38 Devices in the Backplane	167

Illustrations

5.12	(a) Futurebus Trapezoidal Transceiver and (b) Futurebus Termination Circuit	169
6.1	Block Diagram of Typical Data Acquisition System	174
6.2	Interface Block Diagram Within a Typical DAS	175
6.3	Differential Input Multiplexer	176
6.4	(a) Simplified Sample and Hold Circuit and (b) Aperture Time Error	179
6.5	(a) "Black Box" Representation of an A/D Converter and (b) Theoretical Transfer Function of an A/D Converter (first three LSBs only)	180
6.6	(a) "Black Box" Representation of a D/A Converter and (b) Theoretical Transfer Function of a D/A Converter (first three LSBs only)	182
6.7	A Microcomputer-Operated Process Control System	186
6.8	Basic Transducer Concept	187
6.9	(a) Thermocouple to DAS Connections and (b) Circuit Supplying Ambient Temperature Data	189
6.10	V/F Converter Link Using the AD537 Device	191
6.11	F/V Converter as Motor Speed Control	191
6.12	Block Diagram of Single-Line and Data Transmission Link	193
6.13	Alternative Techniques for Voltage Isolation and Analog Data Transmission from Remote Transducers, Using V/F and F/V Converters	194
6.14	Effects of Hostile Environments on Data Acquisition Systems	195
6.15	Piezoelectric Accelerometer Interface	196
6.16	Input Amplifier Common-Mode Problems	197

6.17	Relationship Between Successive Approximations A/D Converter and Op Amp That is the Source of the Input Signal	199
6.18	Bypassing Power Supplies for Virtual-Ground Applications. Arrows Show Unbypassed Current Flow	200
6.19	Decoupling Negative Supply Optimized for "Grounded" Load	201
6.20	This connection minimizes common impedance between analog and digital (including converter digital currents).	201
6.21	Standard Operating Temperature Ranges for Data Converters	203
7.1	Representative Instrumentation Amplifier Circuits: (a) Simple Single Amplifier and (b) Two Followers and Adjustable Gain Added to Circuit in (a)	209
7.2	(a) Model for CMRR Calculation and (b) External Circuitry for Calculation of Effective CMRR	212
7.3	Comparison of Grounding Techniques for Conventional and Isolation Amplifiers	214
7.4	Reduction of Stray Capacitance in Isolation Amplifier Circuit	214
7.5	Power-Up Reset Circuit (dotted line) Added to Provide Internal Initialization	216
7.6	(a) Correct and (b) Incorrect Grounding of Twisted Pairs in an Edge-Connector Fixture	218
7.7	Noise Test Module for Measuring Worst-Case Crosstalk and Ground Noise and Obtaining a Noise Profile of the Test Fixture	217
7.8	Spectrum Analyzer/EMI Receiver System That Automates FCC, VDE and MIL-STD Emission Measurements	223
7.9	A Typical Automatic Test System Using GPIB Controller	224

Illustrations

7.10	Conductive RFI Emission Testing	229
7.11	Peak/Quasi-Peak Detector	230
7.12	Test Setup for Open-Field Radiated EMI Measurements	231
8.1	Basic Types of Optical Isolators: (a) LED-Photodiode, (b) Phototransistor With or Without Base Terminal, (c) LED-Photo-SCR and (d) LED-Photo-Darlington	237
8.2	Digital Interface Isolation Using Optical Isolators in (a) Unidirectional and (b) Bidirectional Data Transmission	238
8.3	Linear Optocoupler Circuit (CTR = Current-Transfer-Ratio and G = Gain)	239
8.4	Schematic of a Standard Laser Power Supply Using Isolation Transformers and Optocoupler	240
8.5	Typical Fiber Optic Link	241
8.6	Types of Optical Fibers—Schematic Representation of Cross Sections, Index of Refraction Distributions of Optical Ray Paths	243
8.7	Ray Paths in a Multimode-Step Index Fiber Optic	243
8.8	Use of Optical Sensors and Single Optical-Fiber Collection Line in a Windmill Energy System	246
8.9	Representative Types of Fiber Optic Connectors	248
8.10	Average Optical Transmission Loss vs. Total Radiation Loss for Fiber Optics	249
8.11	Spectral Compatibility of Emitters, Detectors and Low-Loss Filters	250
8.12	LED Interface Driving Circuits: (a) Series Driven LED and (b) Shunt Driven LED	252
8.13	Block Diagram of Laser Driven Circuit	253

8.14	Optical Receiver Amplifier Circuitry: (a) Bootstrap Interface and (b) Transimpedance Interface	254
8.15	Data Interface With Snap-In Fiber Optic Links: (a) RS-232 Data Interface and (b) RS-422 Data Interface	255
8.16	Multiple Series Driven Transmitters and Wired—OR Receivers in a Three Node Fully Connected Network	256
8.17	A Two-Way Infrared Communication System for Automated Factory Applications	258
8.18	Relative Spectral Energy Distribution for Various Kinds of Sources and PIN Responsivity	260
9.1	Unconstrained Topology. In This Configuration, Each Mode Receiving a Message Must Make a Routing Decision	267
9.2	Representative Schemes of Constrained Topologies: (a) Bus, (b) Star and (c) Ring	268
9.3	Packet Frames for (a) CSMA/CD and (b) Token Passing	270
9.4	Large Ethernet System With Three Coaxial Segments	273
9.5	Wiring Concentrator and Station Interconnects in LAN Ring Configuration Where Shielded Twisted Pairs Are Used	274
9.6	A Fiber Optic Ring System	275
9.7	Block Diagram of IR Drop Configuration for LAN Wireless Applications	276
9.8	Value of the Installed Base of Local Nodes by Medium 1982-1990	278
9.9	Fiber Optic/Coax Digital Hybrid Link	278
9.10	ISO Open Systems Interconnection Model	281
9.11	Major Types of Local Area Networking	282

9.12	EMI Contributions of Power Supply, Digital Equipment and Interfacing Cables to "Antenna Farm" in a LAN System	283
9.13	Shielding in an Ultra-Isolation Transformer, Which Isolates Noise Bidirectionally	284
9.14	Coaxial Transceiver Interface Chip (DP8392) Isolation With Transformers	288
9.15	Signal Isolation in Serial Network Interface Chip (DP8391) by Differential Line Drivers—Line Receivers	289
9.16	Simple Office Network With Individual Test Modules. Each Test Module Recognizes its own Address When Instructed, Then Accepts Commands to Perform its Assigned Functions	291
9.17	Ring Reconfiguration with Alternate Ring. A fault on the ring between two wiring concentrators can be bypassed by wrapping the connections within the wiring concentrators, thereby using the alternate path and maintaining the same node ordering that existed before the fault occurred.	294
10.1	Updated ANSI Standard for RF and Microwave Radiation Exposure	301
10.2	Diagram of RF and Microwave Limits Proposed by NRPB (UK)	302
10.3	Penetration Depth of Microwave Power in Human Tissues With Frequency	303
10.4	Two-Reactance Matching Network	305
10.5	(a) Network With Lumped Constants, (b) Network With a Transmission Line and (c) Bandwidth of L-Section for Both Cases With Transformation Ratio n = 10	307
10.6	Simple Balun: (a) Coaxial Design and (b) Equivalent Circuit (Line A: Z_A, θ_A; Line B: Z_B, θ_B; $Z_p = jZ_B\tan\theta_B$, $Z_p >> R$)	308

10.7	Symmetrical Coaxial Balun: (a) Coaxial Design and (b) Equivalent Circuit (Line A: Z_A, θ_A; Line B: Z_B, θ_B; Line C: Z_C, θ_C)	309
10.8	(a) Two-Section Coaxial Balun and (b) Equivalent Circuit	311
10.9	Microwave Transmission System for LANs	316

Tables

Table	Title	Page
1.1	Frequency Ranges and Band Designations	9
1.2	FCC's Radiated and Conducted Emission Limits	16
1.3	VDE's Radiated Limits	17
1.4	VDE's Conducted Limits	17
1.5	CISPR's Limits for Conducted Interference	19
1.6	CISPR's Limits for Radiated Interference	19
2.1	Key Parameters of TTL Logic	30
2.2	Characteristics of High-Speed ECL Devices	32
2.3	Performance Comparison of High-Speed CMOS With Several Other Logic Families	36
2.4	Maximum Operating Frequency of Representative Devices from Various Logic Families	40
2.5	General ESD Damage Threshold Voltages of Semiconductors	41
2.6	Radiation Hardness—Typical Guidelines	42
2.7	EMI Design Recommendations for Microcircuits	55

9.4	Comparison of LAN Alternative Transmission Media	279
9.5	Representative Line Drivers That Meet EIA Standards	286
9.6	Representative Line Receivers That Meet EIA Standards	287
9.7	Cable Shield Qualities	290
10.1	Basic Elements of the RF/Microwave Interference Problem	313
10.2	Maximum Allowable Field Strength for Nonlicensed Operation	314
10.3	The Major Aerospace EMI Specifications	318
10.4	Basic Comparison of MIL-STD and RTCA Tests	319

Tables

2.8	Input and Output Characteristics of Various Logic Families	59
3.1	Characteristic Impedances for Various Lines	80
3.2	Parameters for Principal EIA Standards	94
4.1	Standards from a Common Carrier's Perspective	106, 107
4.2	International Modem Standards	109
4.3	Touch Technology Comparison	114
4.4	CMOS 80C86 Family Peripheral Support Chips	120
4.5	Peripheral Interface Compatibility	121
5.1	STD Bus Logic Signal Characteristics	142
5.2	The IEEE 488 Lines	145
5.3	Summary of VME Bus Characteristics	154
5.4	Comparison of Multibus and VME Bus Characteristics	156
5.5	Bus Comparison Chart	160, 161
5.6	Parameters of Some of the Leading Buses	162
7.1	Characteristic Requirements for Switching Performance	225
7.2	Computing Device Requirements (47 CFR 15.801—15.840)	227
8.1	Summary of Typical Optical Fiber Characteristics	244
8.2	LEDs vs. Lasers in Fiber Optic Systems	251
8.3	Comparison of Main Characteristics of PIN and APD Detectors	254
9.1	Comparison of Constrained Topologies	269
9.2	CSMA/CD vs. Token Passing Comparison	271
9.3	Transmission Modes Characteristics	272

Abbreviations and Symbols

ac	Alternating Current
A/D	Analog-to-Digital
AFR	Audio Frequency Rectification
ALS	Advanced Low-Power Schottky
AM	Amplitude Modulation
AND	Logic Gate "AND"
ANSI	American National Standards Institute
APD	Avalanche Photo-Diode
ARINC	Aeronautical Radio, Inc.
AS	Advanced Schottky
ATE	Automatic Test Equipment
BEP	Burst Error Processor
BER	Bit Error Rate
BW	Bandwidth
β	Transistor Current Gain
CB	Citizens Band (radio)
CBEMA	Computer Business Equipment Manufacturers Association
CCITT	International Consultative Committee for Telephone and Telegraph
CD	Carrier Detect
CE	Conducted Emission
CISPR	Comite' International Special des Perturbations Radioelectriques
CM	Common-Mode
CMOS	Complementary MOS
CMRR	Common-Mode Rejection Ratio

Abbreviations

CPU	Central Processing Unit	
CRT	Cathode Ray Tube	
CS	Conducted Susceptibility	
CSMA/CD	Carrier Sense Multiple Access/Collision Detection	
CTI	Coax Transceiver Interface	
CW	Continuous Wave	
D/A	Digital-to-Analog	
DAS	Data Acquisition System	
dB	Decibel	
dc	Direct Current	
DCE	Data Circuit Equipment	
DFET	Depletion-Mode FET	
DMA	Direct Memory Access	
DMI	Digital Multiplexed Interface	
DPSK	Differential Phase Shift Keying	
DTL	Diode-Transistor Logic	
DTS	Digital Transmission System	
DUT	Device Under Test	
ECL	Emitter-Coupled Logic	
EHF	Extra High Frequency	
EIA	Electronic Industries Association	
EL	Electroluminescent	
EM	Electromagnetic	
EMC	Electromagnetic Compatibility	
EMF	Electromotive Force	
EMI	Electromagnetic Interference	
EMP	Electromagnetic Pulse	
EMR	Electromagnetic Radiation	
ENFET	Enhancement-Mode FET	
EOS	Electrical Overstress	
ESD	Electrostatic Discharge	
ESDI	Enhanced Small Device Interface	
EUT	Equipment Under Test	
FCC	Federal Communications Commission	
FDM	Frequency Division Multiplexing	
FET	Field Effect Transistor	
FIFO	First-In-First-Out	
FOV	Field-Of-View	
FSK	Frequency Shift Keying	
F/V	Frequency-to-Voltage	

GaAs	Gallium Arsenide
Gbit	Gigabit (10^9 bits)
Gbps	Gigabits per second
GHz	Gigahertz (10^9 Hertz)
GPIB	General Purpose Interface Bus
HCMOS	High-Speed CMOS
HCTMOS	High-Speed TTL Compatible CMOS
HEMT	High Electron Mobility Transistor
HF	High Frequency
HMOS	High-Performance MOS
HNED	Hybrid Nuclear Event Detector
HPIB	Hewlett Packard Interface Bus
HSCMOS	High-Speed Complementary MOS
HSL	High-Speed Logic
HTL	High-Threshold Logic
Hz	Hertz
IC	Integrated Circuit
ICEMOS	Ion-Implanted Complementary Enhanced MOS
IEC	International Electrotechnical Commission
IEEE	Institute of Electrical and Electronics Engineers
iLBX	Local Bus Extension (i = Intel trademark)
I/O	Input/Output
IPI	Intelligent Peripheral Interface
iPSB	Parallel System Bus (i = Intel trademark)
IR	Infrared
iSBX	Parallel System Bus (i = Intel trademark)
ISDN	Integrated Services Digital Network
ISM	Industrial, Scientific and Medical
ISO	International Standards Organization
iSSB	Serial System Bus (i = Intel trademark)
kbit	kilobit
kbps	kilobits per second
kBps	kilobytes per second
kHz	kiloHertz (10^3 Hertz)
LAN	Local Area Network
LCD	Liquid Crystal Display
LED	Light-Emitting Diode
LF	Low Frequency

Abbreviations

LISN	Line-Impedance Stabilization Network
ln	Natural log ($\ln_e$)
log	Common log ($\log_{10}$)
LPS	Linear Power Supply
LS	Low-Power Schottky
LSB	Least Significant Bit
LSI	Large Scale Integration
LSTTL	Low-Power Schottky TTL
mA	Milliamp
MAU	Medium Access Unit
MB	Megabyte
Mbit	Megabit (10^6 bits)
Mbps	Megabits per second
MECL	Motorola Emitter-Coupled Logic
MESFET	Metal Semiconductor FET
MF	Medium Frequency
MHTL	Motorola High Threshold Logic
MHz	Megahertz (10^6 Hertz)
MOS	Metal-Oxide Semiconductor
MSB	Most Significant Bit
mV	Millivolt
μS	Microsecond (10^{-6} second)
μV	Microvolt (10^{-6} volt)
NAND	Logic Gate "NAND"
NED	Nuclear Event Detector
NIC	Network Interface Controller
NIU	Network Interface Unit
NMOS	N-Channel MOS
ns	nanosecond (10^{-9} second)
OEM	Original Equipment Manufacturer
OR	Logic Gate "OR"
OSHA	Occupational Safety and Health Administration
OSI	Open Systems Interconnection
PBX	Private Branch Exchange
PC	Personal Computer
PCB	Printed Circuit Board
PGA	Programmable Gain Amplifier
PIN	Positive-intrinsic-negative

Abbreviations

QAM	Quadrature Amplitude Modulation
RAM	Random Access Memory
RD	Receive Data
RE	Radiated Emission
RF	Radio Frequency
RFI	Radio-Frequency Interference
RFPG	Radio-Frequency Protection Guide
RL	Remote Loopback
ROM	Read-Only Memory
RS	Radiated Susceptibility
RTCA	Radio Technical Commission on Aeronautics
RTL	Resistor-Transistor Logic
R/W	Read/Write
SAE	Society of Automotive Engineers
SAR	Specific Absorption Rate
SASI	Shugart Associates Systems Interface
SBI	Streamer Buffered Interface
SCR	Silicon Controlled Rectifier
SCSI	Small Computer System Interface
SD	Send Data
SDFL	Shottky-Diode FET Logic
SEU	Single Even Upset
S/H	Sample and Hold
SHF	Super High Frequency
SMD	Storage Module Drive
S/N	Signal-to-Noise
SNI	Serial Network Interface
SSI	Small Scale Integration
TEM	Transverse Electromagnetic
TM	Test Mode
TP	Twisted Pair
TSL	Tri-State Logic
T^2L	Transistor-Transistor Logic
TTL	Transistor-Transistor Logic
UHF	Ultra High Frequency
UUT	Unit Under Test

Abbreviations

VDE	Verband Deutscher Electrotechniker
V/F	Voltage-to-Frequency
VHF	Very High Frequency
VHSIC	Very High Speed Integrated Circuit
VLF	Very Low Frequency
VLSI	Very Large Scale Integration
VSWR	Voltage Standing Wave Ratio
WDM	Wavelength Division Multiplexing
ZIF	Zero Insertion Force
Z_o	Characteristic Impedance

Chapter 1
An Introduction to the Interface Domain and EMC Problems

Interface is a shared boundary between system elements defined by common physical interconnection characteristics, signal characteristics and meanings of interchanged signals. As such, it is subjected to interference problems as any other element in the system which it interconnects.

1.1 Introduction

It is well known that an electronic equipment produces electromagnetic fields which may cause interference to other equipment or installations. However, interference is not the only problem caused by electromagnetic radiation. It is possible, in some cases, to obtain information on the signals used inside the equipment when the radiation is picked up, usually from an interface section, and the received signals are decoded.

In electronics, the meaning interference security is covered by the field of electromagnetic compatibility, i.e., the ability of electronic (or electrical) devices or systems to operate in their intended electromagnetic environment at designed levels of efficiency.

At present, references to the various facets of EMC performance of devices, equipment and systems are dispersed throughout the literature, and designers in the field must obtain a large part of their data through manufacturers' component data sheets and application notes. In this book, a coherent approach is developed by applying the information available in the numerous publications to solve potential interface EMC problems.

1.2 Interface Basics

Interface circuits and devices are used whenever data must be transmitted on demand, held temporarily or for a long time, power-amplified, level-shifted, read or extracted from a noisy bus, inverted or otherwise operated upon in some relatively simple electrical way. Interface circuits play the role of "glue" which holds systems together; they are *means* rather than ends in themselves.[1]

1.2.1 Some Useful Definitions[2-4]

1. *Interface* is a shared boundary between system elements defined by common physical interconnection characteristics, signal characterstics and meanings of interchanged signals.
2. *Interface Device* is a device that meets the interface specifications on one side of an interface. The term is usually applied to a device through which a system or equipment (Fig. 1.1) works to meet interface specifications.
3. *Interactive Operation* is an on-line operation where there is a give-and-take between person and machine. Also called *conversational* or *friendly* mode. User presents problem to computer, gets results, asks for variation or amplification of results, gets immediate answer, etc.

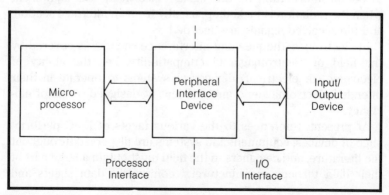

Figure 1.1—Processor-to-I/O Interface

Specifications

4. *Interface Specification* is a set of technical requirements that must be met at an interface.
5. *Physical Interface* defines all interchanges between two interconnected equipment by covering essentially four characteristics:
   a. *Mechanical* (connector, size, etc.)
   b. *Electrical* (voltage, impedance, signal rise time, etc.)
   c. *Functional* (functions, meanings, mandatory subset, etc.)
   d. *Procedural* (sequencing, timing, testing, etc.)
6. *Interface EIA Standard* is a standardized method adopted by the *Electronic Industries Association* to insure uniformity of interface between data communication equipment and data processing terminal equipment and has been generally accepted by most manufacturers of data transmission and business equipment.
7. *Interface CCITT Standard* is a standardized method developed by the *International Consultative Committee for Telephone and Telegraph* and provided as a standard recommendation.

1.2.2 Understanding and Using Interface Specifications[4,5]

Figure 1.2 is a general illustration of connection of data transmission in a communications-served data processing system. Transfer of data between a computer and a public network, or in-plant installation, is carried out by a *Data Terminal Equipment*

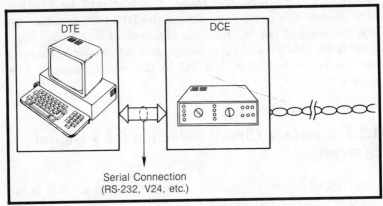

Figure 1.2—General Illustration of Connection of Data Transmission in a Communications-Served Data Processing System

Introduction

(DTE) which is part of the computer system and *Data Circuit Terminal Equipment (DCE)* popularly called a *modem*, which is connected to a transmission line.

An example of the kind of signal exchange commonly handled by a DTE/DCE interface would be as follows: the terminal (DTE), if ready, (power on, paper in printer, etc.) must tell the DCE it is ready. With today's technology, it does this by activating an interface circuit called *Data Terminal Relay*. This particular circuit has this function as its single purpose, and the absence or presence of a signal or energy of a certain level on this circuit can mean only one of two things—that it is ready or it is not ready. Traditional interface standards incorporate this concept whereby each function requires a separate electrical interchange circuit.

The most commonly known interface standard in the United States is EIA RS-232-C; its international equivalent is the CCITT V.24. EIA RS-232-C defines 20 specific functions and, therefore, has 20 interchange circuits. While the physical connector of EIA RS-232-C contains 25 pins, not all of the remaining five are available for use today. From an electrical viewpoint, each EIA RS-232-C interchange circuit consists of a single wire or lead and a ground lead which is shared by all circuits. Interfaces which use the common (shared) ground approach are said to contain *unbalanced* interchanged circuits or, in more common terminology, are simply referred to as *unbalanced interfaces*. Where each interface circuit has its own ground lead which it does not share, the circuit is said to be *balanced*. One of the objectionable features of unbalanced circuits is that the physical DTE/DCE distance separation is limited. The electrical characteristics used in RS-232-C also limit the speed of transfer of information, expressed in bits per second, across the DTE/DCE interface. As will be pointed out in the next chapters of this book, these limitations, plus the interest in both higher data transmission rates and increased functions, has led to the development of new interfaces.

1.2.3 Interface Circuit Selection for a Digital System[1]

According to Ref. 1, two major trends currently evident in the world of interface circuits are:

1. The emergence of an orderly, matrix-like approach to interface products, so that taken all together they form an array rather than simply a mixed group of assorted types and
2. A strong emphasis on increasing the number of data bits which can be handled or accommodated by a single interface circuit package.

In the *real world*, a digital-logic designer does not set out deliberately to use some particular interface circuit whose properties have been carefully learned. Rather, it all starts with some little line between two blocks on the preliminary system block diagram. However, sooner or later the designer will need an interface circuit and will have to go through a several-stage decision process to determine what interface circuit is needed to actually implement that little line before the block diagram can turn into a system.

A *top-down* design approach, as shown in Fig. 1.3, was not the best solution for the designer until recently, simply because desperately needed circuits did not exist. A few years ago, only a quasi-random subset of all the obviously possible variations of the basic interface parts had reached full production status, so that they could be bought and plugged in. The helpless designer just had to memorize what that subset was and do the design *bottom-up* from there. Today, enough of the possible interface parts, which a designer might want to buy, now exist (or will exist shortly) so that the kind of *top-down* thought process shown in Fig. 1.3 will be an efficient one.

Introduction

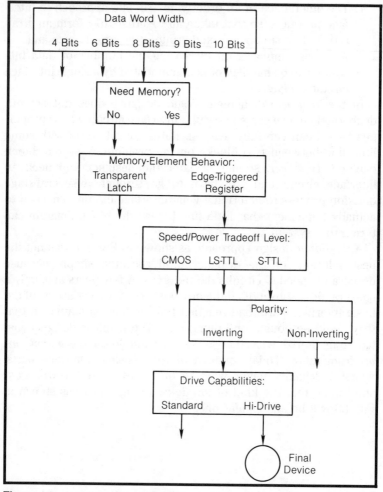

Figure 1.3—Interface-Circuit-Selection Decision Tree (Adapted from Ref. 1)

1.2.4 The Importance of Interfacing and the EMI Problem

Many times, interfaces are the least items to be considered in the race of new technology, or they may be oversimplified during the initial design concept of a system. However, interfaces are vital to the overall system's existence, since:[6]

1. Interfacing means connecting two parts with a third media such as transistor, converter or microprocessor.

2. Interfacing processors with the real world necessitates circuits for translating real world inputs-to-digital data and digital data-to-real world outputs.
3. As digital processing hardware shrinks in size and cost, the need for refined interface circuits increases.
4. The problem is not only to interface, but to interface efficiently, cost-effectively; without burdening the Central Processor Unit (CPU) and its directly associated circuitry.
5. To interface efficiently, the designer must thoroughly understand what must interface with: a) the digital *processor*, and b) the *real* world.
6. The designer must also understand how these interfaces will change over the next few years: e.g., digital to monolithic, Large Scale Integration (LSI) to Very Large Scale Integration (VSLI), silicon to GaAs, etc.

Since many interfaces have to work in harmony with the real world, this "interface" term should be further clarified:[6]

1. It is analog; natural parameters (pressures, flow, speed, etc.) change continuously in infinitely small increments.
2. It is becoming more and more digital; digital sensors, digital communication, thinking/reasoning digitally; brain generates pulses.
3. It is man; many control systems have man in the loop: pilot, driver, operator of word processor, etc.
4. Often it is a combination of all of these.

Well-designed interfaces require not only flexibility and compatibility but should be immune to noise and EMI themselves and prevent such undesirable effects from being pumped into a system. Therefore, both the designer and the user must keep in mind that:

1. Interfacing cables may act as an *antenna farm*.
2. Transients, which generate even outside the facility, may travel along the main distribution down to the power feeder and straight through a dedicated line (interface line) to the critical equipment.
3. Monitoring and alarm interfaces may feed back switching transients and other unwanted signals.
4. Incompatibility in logic family interfaces may cause electrical transients and other EMI effects.
5. In an interconnecting subsystem composed of cable, connector and backshell accessories, the connector is the most susceptible to EMI.

6. Sometimes a material's intrinsic shielding effectiveness is of less concern than is the leakage caused by shield discontinuities such as seams, holes or improper terminations.
7. In a bus system interface, interference problems can occur in many areas of the system, but the principal source is backplane signaling (board-to-board interconnection).
8. In data acquisition systems, an equipment located at a remote site is especially vulnerable to electrical and other hazards. Remote inputs should be protected from overvoltage to prevent destruction of both the equipment and its interface.
9. Test fixture interfaces can be the most critical part in a test installation regarding noise generation, conduction and susceptibility.
10. Optical isolators, fiber optic links and infrared (IR) transmission equipment are among those interfaces which present the least EMI problems.

1.3 Review of Interference Problems

Interference signals originate in the interference source which get into a circuit or system in undesirable ways and cause malfunctioning when the *interference limit* or *redundance* is exceeded. Most of these undesirable signals enter via an interface, be it a cable, a device or an equipment. This section, which starts with the definition of terms, is an overview of interference problems and the challenges facing the engineering community with respect to these problems.

1.3.1 Definition of EMC-Related Terms[7-12]

In electronics, the meaning of *interference security* is covered by the field of *Electromagnetic Compatibility (EMC)*. This field grows in importance in proportion to the influence electronics has in daily life, in all sectors. Below, some definitions are provided regarding interference terms.
1. *Electromagnetic Compatibility (EMC):* The capability of electronic (or electrical) devices, equipments or systems to operate in their intended operational electromagnetic environment at designed levels of efficiency.

2. *Electromagnetic Interference (EMI):* The impairment of a wanted electromagnetic signal by an electromagnetics disturbance.
3. *Radio-Frequency Interference (RFI):* Interference from sources of energy where the interfering signals fall on or near all major critical frequencies of the receiver. Various frequency ranges and band number designations are shown in Table 1.1.

To determine the sources of RFI a distinction should be made between man-made sources and natural sources. *Natural sources* of RFI include: *Johnson noise,* often called *white noise*; noise produced by active circuits like *flicker noise, RF noise* due to delay mechanisms, *shot noise, microphonic noise* and *atmospheric noise*.

Man-made noise can be caused by: *hum* (line frequency influence, usually 50/60 Hz and 100/120 Hz); noise created by

Table 1.1 Frequency Ranges and Band Designations

	Band Number	Band Name	Frequency Range (including lower figure, excluding higher figure)
Band Number Designation	4	VLF, very low frequency	3–30 kHz
	5	LF, low frequency	30–300 kHz
	6	MF, medium frequency	300–3000 kHz
	7	HF, high frequency	3–30 MHz
	8	VHF, very high frequency	30–300 MHz
	9	UHF, ultra high frequency	300–3000 MHz
	10	SHF, super high frequency	3–30 GHz
	11	EHF, extra high frequency	30–300 GHz
	12		300–3000 GHz

	Band	Frequency Range (GHz)
Band Letter Designation	P	0.225–0.39
	J	0.35–0.53
	L	0.39–1.55
	S	1.55–5.2
	C	3.9–6.2
	X	5.2–10.9
	K	10.9–36.0
	Ku	15.35–17.25
	Q	36–46
	V	46–56
	W	56–100

digital signals; *electrical discharges* from electrical motors, electrical switches, corona in transformers, spark plugs and neon lights; *carrier wave (CW)* or *near-CW* signals transmitted from broadcast transmitters or other electronic instrumentation and *pulse-modulated* signals, such as radar.

In examining these signals, it is possible to distinguish two types: narrowband and broadband. *Narrowband* signals, such as man-made CW or near-CW signals, are well-defined in the frequency domain, while *broadband* signals have a very wide and, usually, continuous-looking appearance. Since it has the character of an impulse, discharge noise is a typical broadband noise with very high frequency components.

4. *Electromagnetic Pulse (EMP):* It is caused by electrons ejected from materials by gamma-rays and x-rays emitted from nuclear explosions.
5. *Electrostatic Discharge (ESD):* Is a transfer of electrostatic charge between bodies at different electrostatic potentials caused by direct contact, or induced by an electrostatic field.

1.3.2 Coupling Between Interference Source and Receptor

Every interference case contains three essentials: an EMI source, a receptor and a coupling path between the two. Typical *EMI sources* include transmitters for communications, radar telemetry and navigation; local oscillators in receivers; computers and peripherals; motor switches and power lines; fluorescent lights and special lighting lamps; automobile engine ignitions; diathermy and dielectric heaters; arc welders; electrostatic dischargers and such natural sources as atmospheric noise, lightning and galactic noise. *EMI receptors* include receivers for radar, telemetry and navigation; measuring instruments; computers, sensitive indicators and relays and communication and ordnance.[13]

Coupling path or *coupling element* include antennas; spacing between conductors, shielding and absorptive materials; filters and cables (mutual inductance and capacitance) and ground connections and power lines. The coupling between the source and the receptor takes place by radiation, induction and conduction. Radiation is applicable when the source is at least several wavelengths away from the receptor—a far-field case. Induction relates to a near-field configuration in which the source is only a fraction of a wavelength away. Conduction occurs when the source and receptor are connected by hand wires or metals. However, there is a tendency to use the term *radiative EMI* in both far- and near-field situations.

Radiation depends on the coupling or leakage from the internal source to the outer case, or vice versa (external currents), and on the radiation characteristics of the cabinet. This situation is illustrated in Fig. 1.4. *Electromagnetic Radiation (EMR)*, which is incident upon a system outer enclosure (skin), is coupled through apertures in the skin to the system interior. The interior electromagnetic fields induce RF currents and voltages on the system cables. The RF currents and voltages are conducted to integrated circuits (ICs) located inside the electronic equipment and can cause interference problems to them.[14]

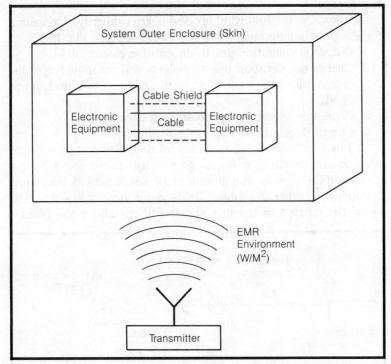

Figure 1.4—EMR Incident upon System Is Coupled through Apertures to System Interior

1.3.3 EMC Performance of Electronic Cables

An electronic cable is one of the main sources, as well as one of the victims, of electromagnetic interference. Being an integral part

Introduction

of electronic devices and computer and communication networks, it affects their compatibility with other equipment and their susceptibility to natural electromagnetic phenomena. Hence, a general concept of the electronic cable, as well as other components, EMC performance can be defined as a measure of their input to the undesirable emissions and susceptibility of the system incorporating these components. Such definition is valid for any component of any electronic system. Electronic cable EMC performance, as defined, is a system parameter. It depends not only on the cable design and characteristics, but also on the cable interaction with other system elements: connectors, terminal equipment and other cables. It also depends on the line working mode, signal parameters, environment properties and grounding techniques.[15]

In summary, the following processes are primarily responsible for electromagnetic radiation from wires (see also Fig. 1.5):[16]

1. Charge acceleration due to an exciting electric field,
2. Charge deceleration due to reflection from impedance discontinuities such as discrete loads and open ends and abrupt bends,
3. Charge deceleration due to smooth curves and
4. Charge deceleration due to dispersion.

At low frequencies, the most useful model is that of a single wire. In all practical cases, a single-wire transmission line is located in the vicinity of a ground plane which acts as the return conductor. In order to estimate fields in this vicinity, one needs to know the current in the line and the height above the ground

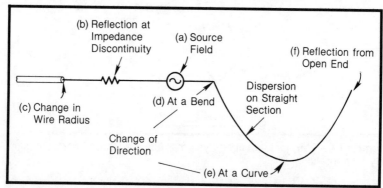

Figure 1.5—Radiation can originate from several places on a wire object. Here they are depicted conceptually and include: (a) the source region; (b) an impedance load; (c) a change in radius; (d) sharp bend (or a junction); (e) a smooth curve and (f) an open end (Reprinted with permission from Ref. 16, © 1980, IEEE).

plane. At low frequencies, the magnetic-field component can be related to the current directly by Ampere's law, which results in a field being attenuated with the inverse of the square of the distance at distances large compared with the height of the line above the ground plane. This model is widely used in computing the fields from open-wire distribution circuits and the coupling to correspondingly open-wire telephone lines.

To avoid coupling from such a conductor, the single-wire situation is usually avoided, often by providing a return current path in a wire close to the original in a parallel line or twisted pair structure so that major cancellation of the field can be obtained. An alternate method is to use a coaxial structure. The electromagnetic principles of all these structures place fundamental limits[17] on the location of susceptible devices or require shielding techniques.

1.3.4 The ESD Problem

Electrostatic Discharge (ESD) can cause degradation, intermittent or catastrophic failures of electronic devices or malfunctioning of electronic circuitry. ESD spark-induced electromagnetic interference can cause temporary malfunction of digital equipment or operating problems. ESD damage to digital devices can also effect timing of circuits. ESD is often the original cause of failures designated as *Electrical Overstress (EOS)* since the typical ESD failure mode is "short." The EOS failure may start out with an ESD from a charged person, or prime static source, resulting in a puncture of a dielectric or a small p-n junction melt within the device. In many cases, when the ESD-damaged device or its higher assembly level (e.g., PCB) is placed in a tester or equipment and normal voltages are applied, high current will flow through the "ESD-caused" short, resulting in an apparent EOS failure. EOS accounts for an average of 30–40 percent of all electronic parts failures. For parts highly sensitive to ESD (e.g., CMOS devices), up to 70 percent of the failures are often categorized as EOS. The actual percentage of EOS failure caused by ESD is unknown.[18]

1.3.5 Challenges Facing the EMC Community

In polling the professionals of the EMC community one and a

Introduction

half years ago, the *Microwave & RF* magazine asked in its first question the following: *What do you consider to be the two or three biggest challenges facing the EMI/RF engineering community?* Mr. Barry Manz, senior editor of the magazine has summarized the responses and put them in the form of a diagram shown in Fig. 1.6.[19]

The respondents cited a lack of test equipment and confusing or non-existent standards as their biggest problems. The electronics industry's reluctance to include susceptibility and suppression

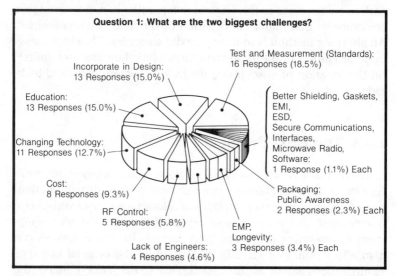

Figure 1.6—Challenges Facing the EMC Community (Survey by Barry Manz, Senior Editor of *Microwave & RF Magazine*, Reprinted by permission, © 1984, Hayden Publishing Co., Inc.)

technology into their products before the first ones come off the line placed a close second, tied with a need for education of management and engineers about the importance of this technology.

The need for education, tying second in the above analysis, is an important factor. It is probably a challenge for engineering educators to start introducing courses relative to EMC subjects. Until then, books like this one can significantly help the EMC engineering community.

1.4 Standard Limits and Measurements

Standards are the essential references for dealing with EMC problems. To ensure compliance, both the designer and the user must understand the national and foreign requirements involved. Unlike the legal profession, EMC engineering has no universally available library where all EMC regulations are tabulated or cross-referenced. Consequently, it is up to the EMC engineer to set up his own library of EMC regulations. Besides attending various EMC courses and seminars and reading EMC-related publications, an easy and rewarding method of keeping abreast of new EMC developments is to participate in one or more of the numerous EMC committees of professional and trade associations that shape and control future regulations.[20] The committees exist at the international and national levels as part of the standardization work of the IEC, CISPR, ANSI, IEEE, SAE, CBEMA, EIA, etc. Most of these committees are open to the EMC engineer interested in furthering the profession.

In the United States, the Federal Communications Commission (FCC), more than any other federal agency, closely regulates electronic hardware. Its EMI regulations cover virtually all devices from video games to mainframes. Products marketed in Europe must meet the requirements of Verband Deutscher Electrotechniker (VDE), International Special Committee on Radio Interference (CISPR) or a comparable agency of a particular nation. These requirements are similar to those of the FCC, but they are by no means the same.

1.4.1 FCC Sets the United States Guidelines[21-25]

In recent years in the United States, there have been major changes in the *Federal Communications Commission (FCC)* requirements pertaining to allowable electromagnetic emissions from electronic devices. The changes have come about due to ever-increasing development and marketing of digital electronic devices. The FCC now regulates nearly 100 kinds of equipment. Specific requirements and technical standards cover nearly all kinds of transmitters, VHF/UHF and CB receivers, diathermics, induction and dielectric heaters and other industrial RF devices, medical and industrial ultrasonic equipment, domestic and industrial microwave ovens, domestic induction heaters, broadcast monitors, marine emergency communication equipment, cordless telephones, and personal and commercial computing devices.

Introduction

The FCC, at the present time, requires that certain categories of electronic equipment and devices meet specifications of FCC Docket 20780, which are included in FCC Rules and Regulations Part 15J. One category includes equipment using digital technology with timing signals and pulses at a rate in excess of 10,000 pulses per second. The second is any device or system that generates and uses radio frequency for the purpose of performing data processing functions, such as electronic computations, operations, transformations, recording, filing, sorting, storage, retrieval or transfer.

The rules governing restricted radiation devices are defined and clarified in FCC Part 15, Subpart J. Two sets of standards have been adopted: *Class A* - equipment for use in a commercial/industrial environment and *Class B* - equipment for use in a residential environment. Table 1.2 shows FCC's radiated and conducted emission limits.

Table 1.2 FCC's Radiated and Conducted Emission Limits

	Radiated Limits	
Frequency (MHz)	Class A at 30 m	Class B at 3m
30–88	30 μV/m	100 μV/m
66–216	50 μV/m	150 μV/m
216–1000	70 μV/m	200 μV/m

	Conducted Limits	
Frequency (MHz)	Class A	Class B
0.49– 1.6	1000 μV	250 μV
1.6 –30	3000 μV	250 μV

Note: From 47 CFR Part 15 J

1.4.2 VDE Sets the European Guidelines[24–26]

The *VDE (Verband Deutscher Electrotechniker)* association of German Electrical Engineers has two standards, VDE 0871/6.78 and 0875/6.77, which are in close agreement with recommendations of the CISPR *(Comite' International Special des Perturbations Radioelectriques)*.

Under VDE 0871/6.78 there are three sets of classifications: *Classes A, B,* and *C.* However, the category C equipment listed

European Guidelines

under VDE is not applicable to mass-produced equipment of the type normally dealt with since it requires an emission test of each installation site. Tables 1.3 and 1.4 show VDE radiated and conducted emission limits, respectively.

Table 1.3 VDE's Radiated Limits

Frequency (MHz)	Class A (μV/m at 30m)	Class A (μV/m at 100m)	Class B (μV/m at 10m)	Class B (μV/m at 30m)	Class C (μV/m at 30m)	Class C (μV/m at 100m)	Class C (μV/m at 300m)
0,01–0,15		500(A)		500(A)		500(A)	500(A)
0,15–0,285		50		50		50	200
0,285–0,49		50		50		250	200
0,49–1,605		50		50		50	200
1,605–3,95		50		50		250	200
3,95–30		50		50		50	200
30–41	500		50		500		200
41–68	30		50		30		200
68–87	500		50		500		200
87–107,828	500		50		500(B)		200
107,828–174	500		50		500		200
174–230	30		50		30		200
230–470	500		50		500		200
470–760	180(C)		100		100		200
760–790	Note D		100		100		200
790–1000	Note D		100		500		200

Note A: Recommended limits only
Note B: Recommended limit is 30 μV/m at 30m
Note C: Measured at 10m
Note D: Measured at 10m; 900 μV/m at 760MHz decreasing linearly to 700 μV/m at 1000MHz
Note E: Class A and B measured at a measurement site
Class C measured at an operational site
Note F: From 0871/6.78 ammended 12.82

Table 1.4 VDE's Conducted Limits

Freq (MHz)	Class A Limit (μV)	Class B Limit (μV)	Class C Limit (μV)	(NORA)
0,01–0,15	(Note B)	(Note C)	(Note B)	
0,15–0,50	2000	500	2000	
0,50–30	1000	250	1000	

Note A: Limits in regulations given in dB μV, here translated to μV
Note B: Suggested limit 35,500 μV at 0,01MHz dropping linearly to 3000 μV at 0,15MHz
Note C: Mandatory limit 8900 μV at 0,01MHz dropping linearly to 750 μV at 0,15MHz
Note D: VDE from 0871/6.78 ammended 12.82

Introduction

VDE 0875 covers equipment not intended to contain an RF source operating above 10 kHz. Such equipment includes household appliances, hand-operated tools and milking machines. VDE 0875's emission limitations are often called broadband because the equipment category is defined as including no RF sources in the frequency range covered by the limits. The other VDE specification, 0871, is often referred to as a narrowband specification because it concerns equipment that contains RF sources above 10 kHz. This category includes all digital equipment, switching power supplies and ISM (*Industrial, Scientific* and *Medical*) equipment that use RF energy for some end function—all devices covered by FCC Parts 15 and 18.

1.4.3 CISPR's Recommendations[27]

The *CISPR (Comite' International Special des Perturbations Radio-electriques)* is an international committee sponsored by the *International Electrotechnical Commission (IEC)*. The committee was formed to set up international RFI standards in industrialized nations. The CISPR follows in its recommendations the same classification of equipment as the FCC. However, according to its latest recommendations CISPR/B(CO), Class A equipment is defined as the data processing equipment and electronic office equipment which satisfies the interference allowable limit for Class A equipment but which does not satisfy the allowable limit for Class B. In some countries, restrictions are placed on the sale and use of this kind of equipment.

As Table 1.5 shows, the CISPR sets allowable limits for conducted interference in terms of quasi-peak values and average values of conducted radiation limits, while the FCC (Table 1.2) sets them only in terms of average values. However, taking into consideration the FCC's measurement method, it may be said that the proposed standards by the two organizations are very close.

The CISPR's radiated limits are shown in Table 1.6.

Table 1.5 CISPR's Limits for Conducted Interference

Class A Equipment

Frequency value (MHz)	Limits (dB/μV)	
	Quasi-peak value	Average value
0.15–0.20	83	70
0.20–0.50	79	66
0.50–5	73	60
5–30	79	66

Class B Equipment Requiring Earth Connection

Frequency range (MHz)	Limits (dB/μV)	
	Quasi-peak value	Average value
0.15–0.20	70	57
0.20–0.50	66	53
0.50–5	60	47
5–30	66	53

Class B Equipment Requiring No Earth Connection

Frequency range (MHz)	Limits (dB/μV)	
	Quasi-peak value	Average value
0.15–0.20	66	53
0.20–0.50	60	47
0.5–5	54	41
5–30	60	47

Table 1.6 CISPR's Limits for Radiated Interference

Frequency range (MHz)	Class A equipment	Class B equipment
	Limits at 30m	Limits at 10m
30–88	30 (dB (μV/m))	30 (dB (μV/m))
88–230	35 (dB (μV/m))	35 (dB (μV/m))
230–1000	37 (dB (μV/m))	37 (dB (μV/m))

1.4.4 Interference Measurement Methodology[10]

Usually, two steps are required when establishing electromagnetic compatibility with electronic equipment. The first step is to perform measurements to determine if any undesired signals being radiated from the equipment (radiated EMI) and appearing on the power lines, control lines, alarm lines or data lines of the

Introduction

equipment (conducted EMI) exceed limits set forth by the using agency. The second step involves the exposure of the electronic equipment to selected levels of *electromagnetic (EM)* fields at various frequencies to determine if the equipment can operate adequately in its intended operational environment. Exposing the equipment to EM fields of various strengths is referred to as *susceptibility* or *immunity testing*.

There are several measurement methods available for making EMC/EMI tests depending on the size of the test equipment, the frequency range, the test limits, the types of field to be measured (electric or magnetic), the polarization of the field and the electrical characteristics of the test signal (frequency or time domain). Test setups may include, without being limited to, open sites, transverse electromagnetic (TEM) cells, reverberating chambers and anechoic chambers.

Open-site measurements provide a straightforward approach to evaluating the EMI performance of the electronic *equipment under test (EUT)*. One important parameter to be determined by this method is the site attenuation, or insertion loss, between a source and a receiver. It can be defined as the ratio of the input power to the source antenna (or the voltage at the source signal generator) to the power induced at the load impedance connected to the receiving antenna (or the voltage developed across the load).

Quite often, the need for sensitivity and the high-level testing field requires near-field measurements. This requirement introduces a problem which is related to field uniformity over the test volume occupied by the EUT and interaction effects. This problem, as well as other difficulties, such as antenna bandwidth limitations and antenna separation, may be eliminated or minimized by using *TEM cells*, because they themselves serve as the transducer and thus eliminate the use of antennas.

Because the polarization of the field generated inside a TEM cell is fixed, the radiated emission and susceptibility tests for an EUT using TEM cells requires physical rotations (or different orientations). A relatively new EMC/EMI measurement technique, which does not require EUT rotations, is to utilize *reverberating chambers* to generate an average, uniformly homogeneous and isotropic field within a load region inside a metal enclosure.

EMI Measurement

Microwave *anechoic chambers* are frequently used for a variety of indoor EMC/EMI measurements and antenna calibrations. The prime requirement for an anechoic chamber in general is that a plane-wave field be introduced to simulate a free-space environment over a test volume (inside the chamber) of dimensions sufficient to perform the test.

1.5 References

1. C. Hastings and B. Braftman, *Pick the Right 8-Bit—Or 16-Bit-Interface Part of the Job*, Proceedings of Northcon/82, May 1982, pp. 18–20.
2. C. J. Georgopoulos, *Interface Fundamentals in Microprocessor-Controlled Systems* (Holland: D. Reidel Publishing Co., 1985).
3. J. M. Nye, *Users Guide: Voice & Data Communications Protection Equipment*, NTIA-CR-80-9 (Washington, D.C.: Government Printing Office, December 1980).
4. N. Cowder and J. Merkel, "A Tale of Two Interfaces," *Telecommunications*, April 1982, p. 57.
5. J. Steeman, "RS 232 Interface," *Elektor Electronics*, September 1985, pp. 76–81.
6. H. Schmid, *Interfacing Monolithic Processors With the Real World* (N.Y.: General Electric Company).
7. A. H. Kunz, *Origin and Simulation of Mains Interference*, EMC '81—International IEEE Symposium on EMI Compatibility, Boulder, Co., August 18–20, 1981.
8. W. E. Cracker and W. H. Siemens, *Electromagnetic Compatibility Test Facility for Signal and Communication Systems*, Annual Meeting of Communication & Signal Div., Association of American Railroads, August 19, 1980.
9. F. Jay, *IEEE Standard Dictionary of Electrical and Electronic Terms*, The Institute of Electrical and Electronic Engineers, Inc., N.Y., 1977.
10. M. T. Ma, et al., "A Review of Electromagnetic Compatibility/Interference Measurement Technologies," *Proceedings of the IEEE*, Vol. 73, No. 3, March 1985, pp. 388–411.
11. R. C. Kerns and J. R. Riskin, *ESD Prevention Manual Protecting ICs from Electrostatic Discharges*, Analog Devices, Inc. 1984.
12. G. Sorger, *Measuring RFI—Spectrum Analyzer vs. RFI Receiver*, Communications Engineering International, April 1984, p. 53.
13. D. R. J. White, K. Atkinson, and J. D. M. Osburn, "Taming EMI in Microprocessor Systems," *IEEE Spectrum*, December 1985, pp. 30–37.

References

14. K.N. Chen, et al., *Using Macromodels to Compare RFI Bipolar and FET-Bipolar Operational Amplifiers,* EMC/81—4th Symposium and Technical Exhibition on Electromagnetic Compatibility, Zurich, Switzerland, March 10-12, 1981.
15. A. Tsaliovich, *Defining and Measuring EMC Performance of Electronic Cables,* 1983 IEEE International Symposium on Electromagnetic Compatibility, Crystal City, Va., 1983.
16. E. K. Miller and J. A. Landt, *Direct Time-Domain Techniques for Transient Radiation and Scattering from Wires,* Proceedings of the IEEE, Vol. 68, No. 11, November 1980, pp. 1396–1423.
17. R. M. Showers, et al., *Fundamental Limits on EMC,* Proceedings of the IEEE, Vol. 69, No. 2, February 1981, pp. 183–195.
18. Reliability Sciences Incorporated (RSI), *Electrostatic Discharge (ESD) Control Capabilities,* Arlington, Va.
19. B. Manz, *An Active EMC Community Struggles for Acceptance,* Microwaves & RF, October 1984, p. 67.
20. H. K. Mertel, *Keeping Up With Changes of EMC Rules and Regulations,* EMC/81—International IEEE Symposium on Electromagnetic Compatibility, Boulder, Co., August 18–20, 1981.
21. S. Bernstein and R. Palker, *A Tutorial Overview of Emission Control for FCC Part 15.J Compliance,* EMC Technology, January-March 1983, p. 60.
22. M. Mobley, *FCC and EMC: Origin and Development of the Equipment Authorization Program,* ITEM-1983, p. 152, R&B Enterprises, 1983.
23. I. Straus, *Design Digital Equipment to Meet FCC Standards,* EDN, June 5, 1980, pp. 141–144.
24. M. Head, *Know RF-Emission Regulations Pertaining to Your Designs,* EDN, August 19, 1981, pp. 149–154.
25. C. H. Kuist, *Solving VED/FCC Compliance Problems,* Communications Engineering International, November 1984, p. 49.
26. H. K. Mertel, *The VDE or FTZ RFI Requirements,* ITEM-1983, R&B Enterprises, 1983, p. 49
27. M. Okamura, *Radio Interference and International Standards,* JEE, February 1983, p. 56.

Introduction

General References

- R. A. King, *How Real World Applications Are Being Supported Using Unusual Interfacing Techniques*, Southcon/85 and Mini/Micro Southeast/85, Atlanta, GA, March 5–7, 1985.
- F. D. Williams, *Electromagnetic Interference and the Digital Era, Ham Radio*, September 1984, p. 114.
- D. R. J. White, *EMI Control Methodology & Procedures*, Third Edition, DWCI, Gainesville, VA, 1982.
- M. Mardiguian, *Electrostatic Discharge—Understand, Simulate and Fix ESD Problems*, DWCI, Gainesville, VA, 1982.
- M. N. Yazar, *New Directions for EMC Standards*, 5th Symposium and Technical Exhibition on Electromagnetic Compatibility, Zurich, Switzerland, March 8–10, 1983.
- J. F. Kalbach, *Designer's Guide to Noise Suppression, Digital Design*, January 1982, p. 26.
- N. S. Sahman, et al., *Methodology for Standard Electromagnetic Field Measurements, IEEE Transactions on Instrumentation and Measurements*, Vol. IM-34, No. 4, December 1985, pp. 490–503.

Chapter 2
Logic Elements Interfacing

This chapter deals with the basic *logic families*. The presentation is not only in terms of device operation, but also in terms of performance, capabilities and behavior under interference conditions. The designer must know those qualities in order to turn out a good and reliable electronic product.

2.1 Introduction

After a review of the basic TTL, ECL, CMOS, GaAs logic families and their new versions, various interconnection approaches for minimizing noise and other *interference problems* are examined. It is shown that as circuit speeds increase, wiring rules and system design rules must be adjusted accordingly.

Emphasis is also placed on techniques of *interfacing* different logic families. When connecting digital ICs that belong to the same family, there will rarely be any problems. It is quite a different matter, however, to try to match devices from different logic families. In many cases, extra adapter circuitry is required to ensure proper interfacing. Such circuitry may vary from a simple pull-up resistor to a complete IC package.

This chapter is relatively long since it includes material which is essential to the structure and understanding of most of the subsequent chapters of this book.

Logic Interfacing

2.2 Review of Basic Logic Families

A good knowledge of the characteristics and capabilities of both traditional and new logic families helps in designing and interconnecting devices, circuits and systems with a minimum number of pitfalls. To establish proper interfacing from one family to another, an understanding of the input/output characteristics and noise immunity of each family is necessary.

2.2.1 TTL Family

For many years, *transistor-transistor logic (TTL or T^2L)*, which has displaced *resistor-transistor logic (RTL)* and *diode-transistor logic (DTL)*, has been the primary form of logic circuit used to perform these functions and interface with neighboring circuitry. While many manufacturers are making TTL with the same functional appearance, few TTL ICs have equivalent circuits or performance—even when part numbers are identical. Among the several designs that currently exist, many appear similar to the uninitiated engineer, but small distinctions make the difference between the proper system functioning or not. Such circuit design variations greatly affect the performance, cost and application of TTL devices. A typical TTL NAND gate is shown in Fig. 2.1.

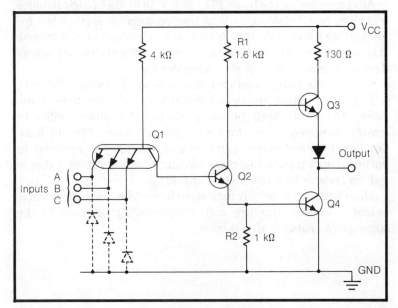

Figure 2.1—Typical TTL NAND Gate

TTL Family

The interface voltage characteristics required at the output and input terminals of saturated logic, as in the TTL family, are shown in Fig. 2.2,[1] where,

V_{CC} = The voltage of power supply,
V_{IL} = The voltage value required for a low-level input voltage that guarantees operation,
V_{IH} = The voltage value required for a high-level input voltage that guarantees operation,
V_{OL} = The guaranteed maximum low-level output voltage of a gate and
V_{OH} = The guaranteed minimum high-level output voltage of a gate.

The difference between the output voltage levels guaranteed for the driving gate and the input voltages required for the driven gate defines the *low-state noise margin* (NM_L) and the *high-state noise margin* (NM_H), i.e.,

$$NM_L = (V_{IL})_{max} - (V_{OL})_{max} \qquad (2.1a)$$
$$NM_H = (V_{OH})_{min} - (V_{IH})_{min} \qquad (2.1b)$$

One of the peculiar drawbacks to TTL circuits is that the outputs of two gates with totem-pole structures cannot be directly connected in what is normally known as a *wired-OR function*.

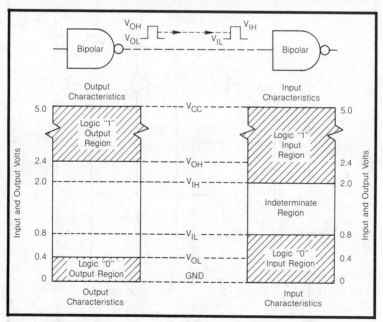

Figure 2.2—Voltage Characteristics Required at the Output and Input of Typical TTL Family Logic

Logic Interfacing

One of the solutions to this problem provides the *tri-state logic (TSL)*. Tri-state logic is essentially TTL with output stages, or input and output stages, that can assume three states. Two states are normal low-impedance TTL "1" and "0" states. The third is high-impedance or *hi-Z* state that allows many tri-state devices to time-share bus lines. These devices have the speed of standard TTL, higher line-drive and noise immunity. By eliminating pull-up resistors, they cut bus delays to a few nanoseconds. A typical tri-state gate structure is shown in Fig. 2.3.

Schottky TTL, a derivative of TTL, is a logic circuit (Fig. 2.4) designed to increase TTL speed in order to approach the speed of emitter-coupled logic (ECL). This logic is fully compatible with other members of the TTL family and equally as easy to use. The configuration for Schottky TTL is basically the same as standard TTL except that Schottky barrier-diode clamped transistors are included in the circuit. Because the Schottky barrier-diode incor-

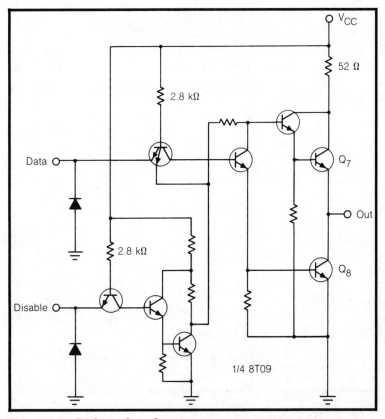

Figure 2.3—Tri-State Gate Structure

porated in a transistor is extremely fast and the transistor is not saturated, switching speed is increased and propagation delays are reduced.

The Schottky line, besides the regular and *low-power Schottky (LS)* devices, now includes *advanced Schottky (AS)* and *advanced low-power Schottky (ALS)* products that provide lower power, faster speed, and dense gate structure.[2]

The original 54/74 family of TTL logic devices provides a convenient speed/power product reference point. A typical TTL gate offers a 10 ns delay time at a power dissipation of 10 mW, giving it a speed/power product of 100 pJ. Another important parameter is the maximum *toggle frequency* under which a flip-flop will operate. Of course, system designers should be very careful in using this figure for system designs since there are always circumstances under which a device will not operate properly at the maximum frequency. It is, however, a useful guidepost. Again, the reference point is the TTL, and the maximum frequency is 35 MHz. Table 2.1 compares the highest performance bipolar technologies available to designers, including 54/74 TTL for reference.[3] Basically, the two primary technologies in existence are *Schottky-clamped* and *gold-doped*. The propagation, speed/power product and maximum operating frequency are listed for each logic family within the two technologies.

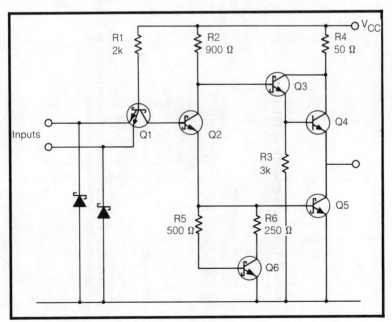

Figure 2.4—Schottky TTL Logic Circuit

Table 2-1. Key Parameters of TTL Logic*

Circuit Technology	Series	Gates			Flipflop Frequency Range
		Propagation Delay Time	Power Dissipation	Speed/Power	
Schottky-clamped	SN54S/74S	3 ns	19 mW	57 pJ	dc to 125 MHz
	SN54LS/74LS	9.5 ns	2 mW	19 pJ	dc to 45 MHz
	SN54AS/74AS	1.5 ns	8 mW	12 pJ	dc to 200 MHz
	SN54ALS/74ALS	4 ns	1.2 mW	4.8 pJ	dc to 70 MHz
	54F/74F	3 ns	4 mW	12 pJ	dc to 175 MHz
Gold-doped	SN54H/74H	6 ns	22 mW	132 pJ	dc to 50 MHz
	SN54/74	10 ns	10 mW	100 pJ	dc to 35 MHz
	SN54L/74L	33 ns	1 mW	33 pJ	dc to 3 MHz

*From "Speed Up, Power Down for Bipolar Logic" by W. T. Greer, Jr., and B. Bailey, p. 162. Copyright by Computer Design, November 1984. All rights reserved. Reprinted by permission.

2.2.2 ECL Basics

The basic gate structure of Fig. 2.5 shows that the natural *emitter coupled logic (ECL)* performs the positive logic function OR/NOR, compared to NAND for TTL gate. V_{CC} is usually considered as system ground in ECL designs. The output voltage swing is generated across R_L and is therefore referred to V_{CC}. Also, any variation in V_{CC} between gates causes a direct reduction in the noise margin, while variations in V_{EE} appear in common mode across the amplifier, thus having a far smaller effect. Most ECL circuits are specified with $V_{EE} = 5.2$ V. The V_{BB} reference voltage is generated on the chip and is more or less temperature-compensated, depending on the device manufacturer, to allow thermal gradients across systems without undue shift in logic threshold. A typical low level is -1.7; a typical high level is -0.9 V.

The guaranteed noise margin is only 125 mV in the high state, although typical noise margins are 200 mV in either state. These margins appear very low to a designer familiar with TTL, but the transmission line techniques generally required at ECL speeds to preserve signal purity also reduce the susceptibility of signals to noise pickup. With careful design, the low noise margins do not present a problem.

The first commercial logic family using ECL technology was the MECL family introduced by Motorola in 1962. With an 8 ns propagation delay and a 30 MHz switching rate, it was the fastest logic available of its time. By 1968, MECL III had been introduced

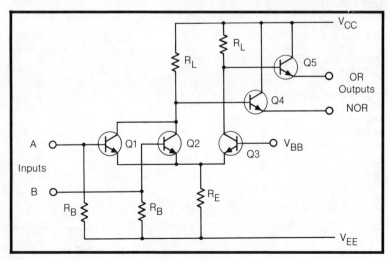

Figure 2.5—Basic ECL Gate Structure

which provided 1ns propagation delays and 300 MHz switching rates. Today, three ECL families predominate. The first is 10K, introduced in 1971. It is not as fast as MECL III, but it uses half the power of MECL III and operates at 2 ns propagation delays. Its 3.5 ms rise and fall times make it easier to design with than MECL III.

The second family, in terms of performance, is 10kH which is basically an improved 10K family. At equal power dissipation to 10K, it operates at twice the speed and is fully compatible with the 10K family.

The third family, F100K, is the fastest full-line ECL family in production today. It operates at 750 ps propagation delays and offers 350 MHz switching rates. The F100K family is a radical departure from the 10K family in several areas, making compatibility difficult. Table 2.2 lists some typical characteristics for high-speed devices of the ECL family.[4]

ECL noise margin is defined in terms of the specification points surrounding the switching threshold.[5] The critical parameters of interest here are those designated with "A" subscript ($V_{OHA\ min}$, $V_{OLA\ max}$, $V_{IHA\ min}$, $V_{ILA\ max}$) in the transfer characteristic curves of Fig. 2.6. Guaranteed noise margin (NM) is defined as follows:

$$NM_L = V_{ILA\ max} - V_{OLA\ max} \qquad (2.2a)$$
$$NM_H = V_{OHA\ min} - V_{IHA\ min} \qquad (2.2b)$$

Table 2.2 Characteristics of High Speed ECL Devices (Adapted from Ref. 4)

Family→ ↓ Feature	10 K	10 KH	MECL III	F100K
Propagation Delay (ms)	2.0	1.0	1.0	0.75
Power (mW)	25	25	60	46
Power-Speed Product (pJ)	50	25	60	44
Rise/Fall Times (ms)	2.0	1.5	0.6	0.8

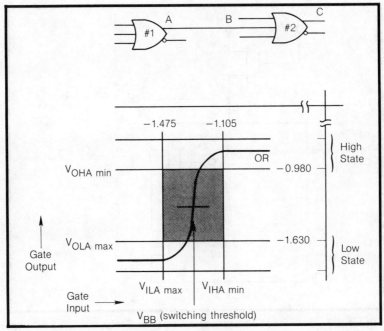

Figure 2.6—Specification Points for Determining Noise Margin in ECL Logic Family

2.2.3 The CMOS Family

The decade of the 1970s was ruled by single-channel *metal-oxide-semiconductor (MOS)* devices. The first MOS products were metal-gate *p-channel MOS* chips which were fabricated with only four key photomasks.[6] Currently, the most popular MOS technology is *n-channel MOS* due to its high packing density and fast switching speeds. To emphasize the high-performance feature, *Intel* prefers to call it *HMOS*. A leading competitor is *complementary MOS (CMOS)* technology which offers faster speed and lower power consumption than circuits implemented with traditional p-MOS and n-MOS technology. Figure 2.7 shows the implementation of a CMOS gate, whereas Fig. 2.8 shows the CMOS input and output characteristics at $V_{DD} = 5$ V.

CMOS devices are designated to switch at a voltage level one-half the power supply voltage in comparison to $+1.5$ V of TTL devices which is not one-half of the supply voltage.[1] Equations

Logic Interfacing

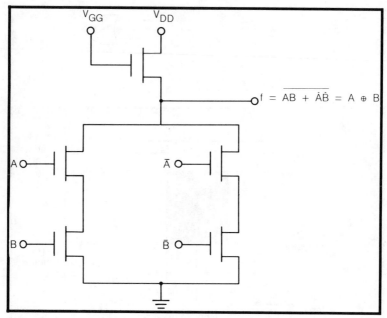

Figure 2.7—Two-Stage Implementation of the Exclusive-OR Function

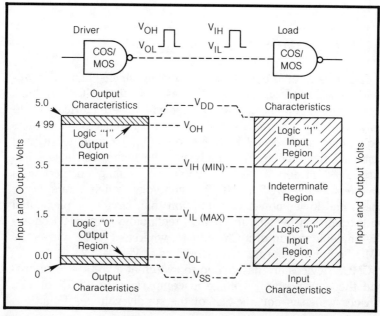

Figure 2.8—Typical CMOS Input and Output Characteristics at a Power Supply Voltage of 5 V

2.1a and 2.1b are also applicable here for calculating the low-state noise margin (NM_L) and high-state noise margin (NM_H) of CMOS devices. Substituting the indicated levels on Fig. 2.8 into the above equations, NM_L = 1.49 V and NM_H = 1.49 V are obtained. However, these noise margin levels will be different for the new CMOS devices that are discussed below.

Traditionally, CMOS logic has been used whenever low power, high noise immunity or wide operating voltage range was important. CMOS logic with a maximum operating frequency of 1 MHz is slow compared to LS TTL, making it unsuitable for use in microprocessor systems and other systems where speed is important. But, improvements in semiconductor processes have resulted in faster devices with more drive capability that are seriously challenging the established LS TTL. Those devices belong to the *high speed logic (HSL) or HSCMOS* family.[7]

The new forms of CMOS logic, called HC and HCT, operate at about the same speed as LS TTL (8 to 10 ns propagation delay time) but they dissipate less than one-tenth the power of the bipolar types. In fact, the HCT family is designed specifically as a pin-for-pin replacement for LS TTL chips in existing systems. When converting form LS TTL to HCT, designers maintain the same levels of performance while reducing power consumption. The HC series, on the other hand, contains many more devices than HCT and is intended for new designs.[8] For the growing number of manpack-type systems requiring extremely low power consumption, portability and battery powered operation, HCCMOS provides all the advantages of its 4000B forerunner—wide operating temperature range, high noise-immunity, very low dissipation—with the added benefit of TTL-like speed capability, as seen from Table 2.3.[9]

It should be mentioned here that a recent development in the metal-gate process has resulted in a product that combines the best of the CMOS and bipolar worlds. Called *ion-implanted complementary enhanced metal oxide semiconductor (ICEMOS)*, it confers upon the advanced high-speed CMOS 54/74 AHCT series the same low propagation delays and high drive capability that characterize the advanced low-power Schottky (ALS) chip family. Actual figures are below 5 ns for average propagation delay and up to 24 mA for drive current.[10]

It is known that for CMOS, ECL or TTL devices, system noise which is great enough can affect the logic integrity. In HC-CMOS using a 4.5 V V_{CC}, typical output levels are ground and V_{CC}, and

Logic Interfacing

Table 2.3 Performance Comparison of High-Speed CMOS with Several Other Logic Families*

Technology‡	Silicon-Gate CMOS	Metal Gate CMOS	STD TTL	Low-Power Schottky TTL	Schottky TTL	Advanced Low-Power Schottky TTL	Advanced Schottky TTL
Device series	SN74HC	4000	SN74	SN74LS	SN74S	SN74ALS	SN74AS
Power dissipation per gate (mW)							
Static	0.0000025	0.001	10	2	19	1	8.5
At 100 kHz	0.17	0.1	10	2	19	1	8.5
Propagation delay time (ns) (C_L = 15 pF)	8	105	10	10	3	4	1.5
Maximum clock frequency (MHz) (C_L = 15 pF)	40	12	35	40	125	70	200
Speed/Power product (pJ) (at 100 kHz)	1.4	11	100	20	57	4	13
Minimum output drive (mA) (V_O = 0.4 V)							
Standard outputs	4	1.6	16	8	20	8	20
High current outputs	6	1.6	48	24	64	24/48	48/64
Fan-out (LS loads)							
Standard outputs	10	4	40	20	50	20	50
High-current outputs	15	4	120	60	160	60/120	120/160
Maximum input current, I_{IL} (mA) (V_I = 0.4 V)	±0.001	−0.001	−1.6	−0.4	−2.0	−0.1	−0.5

‡Family characteristics at 25°C, V_{CC} = 5 V; all values typical unless otherwise noted. This table is provided for broad comparisons only. Parameters for specific devices within a family may vary. For detailed comparisons, please consult the appropriate data book.
***Courtesy of RCA Corporation**

input thresholds are $V_{IH} = 3.15$ V and $V_{IL} = 0.9$ V. These figures yield a noise margin approximately 1,300 mV (logic "1") and 850 mV (logic "0"). LSTTL's immunity is 700 mV and 400 mV, respectively. When using HCT with HC, output logic levels are almost equal to power-supply levels. Therefore, HCT's specified noise margin is approximately 700 mV for a logic "0" and 2.4 V for a logic "1."

2.2.4 GaAs ICs

Gallium Arsenide (GaAs) integrated circuits are a recent development of solid state devices. Since their introduction by Harris Corp. in 1984, they have been used in a wide variety of systems to provide improved accuracy and greater processing capability, as well as broad operating temperature ranges and better radiation tolerances. Although the most obvious applications of GaAs ICs are in doubling or quadrupling the speed of digital signal processing, new applications are appearing. The design improvements currently being realized through use of GaAs ICs include[11] replacing emitter-coupled logic (ECL) products one-for-one in order to remove signal status uncertainty in marginal system designs, expanding the usable bandwidth and throughput available for digital signal processing beyond the level currently available with ECL in systems such as fiber-optic links and high-speed conductor test equipment and improving the accuracy, synchronization and timing of measurement systems.

Replacement of ECL products with GaAs ICs has already produced interesting results. For example, many high-speed ECL circuits can self-oscillate if no input clock or signal is present. In contrast, GaAs ICs using MESFET technology will not self-oscillate. ECL manufacturers, furthermore, specify a guaranteed operating range over a certain bandwidth. That operating range can have holes or frequency dropouts below the specified minimum frequency. GaAs ICs, however, can operate from dc to their maximum operating speed without any holes in frequency coverage.

There are several device choices for high-speed GaAs ICs, each one having certain advantages and disadvantages for logic applications. Three early GaAs logic approaches are shown in Fig. 2.9. *The depletion-mode FET (DFET)* is the most mature, the fastest and the most power-hungry device. The *Schottky diode-FED logic*

Logic Interfacing

(SDFL) claims advantages in circuit density, while *enhancement-mode FET (ENFET)* offers the simplest, densest circuits at lower power consumption but with some sacrifice in speed.[12]

The newest silicon technologies include GaAs *metal semiconductor field effect transistors (MESFET)*, *heterojunction bipolar transistors (HBT)* and *high electron mobility transistors (HEMT)*. The impact of GaAs technology on computer memory performance will be tremendous. It is expected that a broad spectrum of computer, communication and instrumentation systems will achieve new levels of performance through the utilization of commercial GaAs nanosecond RAMs.[13,14]

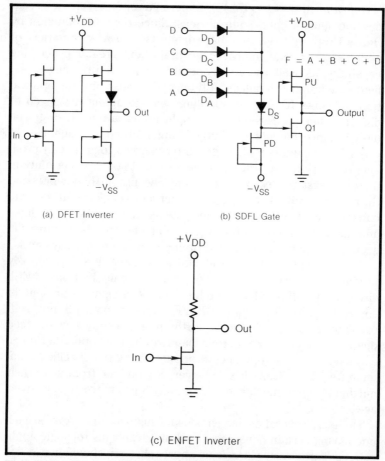

Figure 2.9—Equivalent Circuits of GaAs Logic Gates

2.2.5 Technology Comparison

Traditionally, ECL has been the technology of choice for high-performance systems despite its disadvantages. The main contenders for high-performance logic today are ECL, Schottky TTL NMOS/CMOS and GaAs. Figure 2.10 presents a comparison of those devices with respect to their power dissipation/gate and gate switching speed.[15]

The power dissipation per gate of GaAa ICs is about the same as that of silicon NMOS, but its switching speeds are 10 times faster. Silicon emitter-coupled logic is the fastest existing commercial IC technology, but its power dissipation restricts packing density. Only CMOS has a speed-power product comparable to that of GaAs.

Table 2.4 compares the maximum operating frequency for various logic families, including 54/74 Standard TTL as reference. When manufacturers of high-speed CMOS logic (HCMOS) make comparisons against bipolar technology, the LS family is usually the bipolar representative. It should be noted that recent developments in GaAs technology have shifted the minimum operating frequency to much higher levels than the 3,000 MHz range indicated in Table 2.4.

From Fig. 2.11 note that special attention has been given to

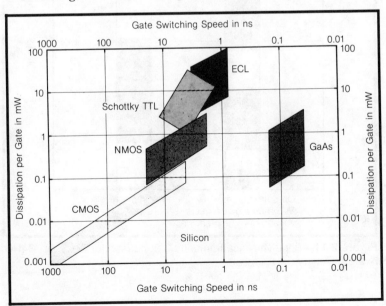

Figure 2.10—Dissipation for Gate vs. Gate Switching Speed for Various Logic Families (© 1983, IEEE[15])

Logic Interfacing

Table 2.4 Maximum Operating Frequency of Representative Devices from Various Logic Families

Logic Family	Series' Representative Device	Maximum Operating Frequency	Cited Reference
TTL STD	SN54/74	35 MHz	3
TTL S	SN54S/74S	125 MHz	3
TTL LS	SN54LS/74LS	45 MHz	3
TTL AS	SN54AS/74AS	200 MHz	3
TTL ALS	SN54ALS/74ALS	70 MHz	3
TTL FAST	54F/74/F	175 MHz	3
CMOS	4000	12 MHz	8
CMOS HC	5N54HC	40 MHz	8
ECL	F100K	325 MHz	38
GaAs	10G011	3000 MHz	38

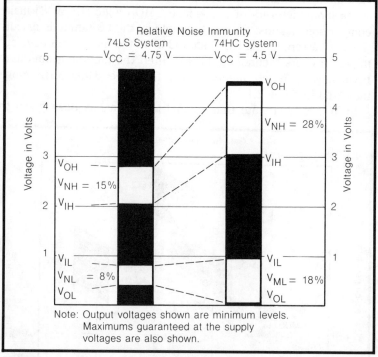

Figure 2.11—Relative Noise Immunity Comparison for 74LS and 74HC System Design

Technology Comparison

noise immunity in the new high-speed CMOS series in comparison to bipolar LS TTL devices. The noise immunity of SN74LS devices is 15 percent (V_{NH}) and 8 percent (V_{NL}) of the minimum supply voltage (4.75 V). SN74HC parts increase this margin to 28 percent and 18 percent respectively (at V_{CC} = 4.5 V).[16-19] Output voltages shown are the min/max guaranteed at the supply voltages listed.

Designers also need to know the sensitivity of those and other semiconductor devices to *electrostatic discharge (ESD)* effects as well as their *transient radiation upset*. Table 2.5[20] shows a brief summary of ESD susceptibility for common technologies and devices. Sensitivity of electronic devices to ESD varies depending on materials and configuration and on the shape of ESD impulse.[21] Some metal-oxide semiconductor (MOS) devices with no protective circuitry can be damaged by as little as 30 V. Therefore, protection networks are necessary to maintain a high level of device reliability.[22-24]

Another important characteristic, especially applicable to military and space systems, is the behavior of IC devices to radiation effects. Table 2.6 lists typical guidelines for radiation hardness and shows the characteristics of ALS, AS and HCMOS in three categories: total dose, neutron bombardment and *single event upset (SEU)*. All three families offer total dose characteristics well

Table 2.5 General ESD Damage Threshold Voltages of Semiconductors*

ESD Voltage Range	Classification	Device
30V to 400V	Most Sensitive	—MOS without protection —Discrete Schottky diodes
400V to 1.5kV	Sensitive	—LSTTL and STTL logic —RF transistors —CD 4000A-series CMOS logic —LSI MOS logic
1.5kV to 4kV	Less Sensitive	—ECL logic —Second generation CD4000B-series CMOS logic —JFET's and some discrete bipolar
4kV to 15kV	Much Less Sensitive	—Some linear bipolar —TTL logic —Some discrete bipolar
15kV and up	Insensitive	—Some linear bipolar —Power semiconductors —Zener diodes

*Reproduced by kind permission from Electronic Engineering, © Morgan Grampian Publishers Ltd., and the author R. Funk.

Logic Interfacing

above the tactical-system typical requirements. The neutron and SEU hardness ratings are also within acceptable limits for tactical systems hardware.[8]

Table 2.6 Radiation Hardness—Typical Guidelines*

Technology	Total Dose (RAD-SI)	Neutron (N/CM²)	Single-event upset SEU (Errors/Bit-Day)
Advanced Low-Power Schottky ALS-Walled Emitter	$>2 \times 10^4$	10^{14}	$10^{-5} - 10^{-6}$
Advanced Schottky (AS)	$>2 \times 10^4$	10^{14}	$10^{-5} - 10^{-6}$
High-Speed CMOS (HC)	2×10^4	$>10^{14}$	$10^{-6} - 10^{-7}$

Source: IEEE NSRE Transactions '80–'83.
Note: Data Sheet Parametric Failure Defines Device Failure.
*From Defense Electronics, © 1984 by EW Communications. Reprinted by permission.

2.3 Interconnection Approaches and Minimization of Interference Problems

Some of the ICs can be connected directly to other chips while others may need special interfacing circuitry. In both cases, designers must be aware of the noise margin and other EMI and ESD effects, as well as the procedure for reducing those effects, when designing equipment or a system. This section deals with the sources of those problems and provides some guidelines to designers for compliance with FCC and other EMI regulations.

2.3.1 Bandwidth Demands of Fast Logic

As the speed of logic families increases, propagation delay times and rise and fall times decrease. Thus, the problem of getting data from one point to another in a high performance system becomes more difficult, often as a direct result of the bandwidth of the interconnection system required to carry data pulses. In Fig. 2.12a, graph plots approximate system interconnection bandwidth for maximum performance as a function of device rise time for popular logic families.[25,26] Links marked N=1 and N=10 plot the equation of bandwidth (BW) required to pass the pulse signal:

$$BW = 0.35N/t_r \qquad (2.3)$$

In the above equation, t_r is the device rise time and N is the highest

Bandwidth Demands

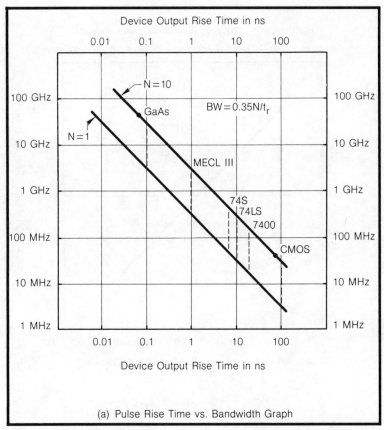

Figure 2.12a—Bandwidth, Pulse Rise Time and Critical Wavelength Plotted Relationships (from "Interconnection System Approaches for Minimizing Data Transmission Problems" by R.K. Southhard, p. 107 and p. 109. Copyright by Computer Design, March 1981. All rights reserved. Reprinted by permission)

harmonic of the propagation frequency to be passed. At $N=1$ the system distorts the pulse rise and fall times and barely passes the leading and trailing edges of the desired pulse. In the normal high performance situation, at $N=10$, the system passes signals easily, with almost no pulse edge distortion. For example, to use 74LS parts to maximum advantage, the system bandwidth must extend from the very high frequency (VHF upper frequency = 30 MHz) range to the ultrahigh frequency range (UHF 300-3,000 MHz) encountered mostly by radio frequency (RF) circuit designers.

System performance can be degraded as the wavelength of the highest frequency being transferred approaches the transmission path length. After the data transmission path exceeds a significant

Logic Interfacing

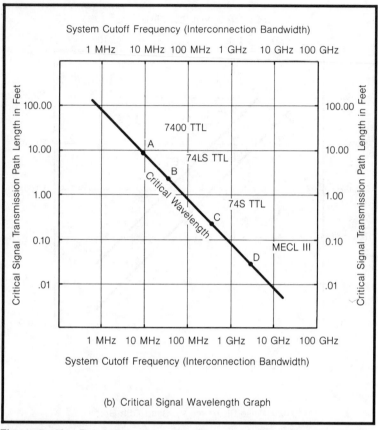

Figure 2.12b—Bandwidth, Pulse Rise Time and Critical Wavelength Plotted Relationships (from "Interconnection System Approaches for Minimizing Data Transmission Problems" by R.K. Southhard, p. 107 and p. 109. Copyright by Computer Design, March 1981. All rights reserved. Reprinted by permission)

fraction of a wavelength, a transmission line approach must be introduced with its attendant factors including characteristic impedance, attenuation and reflection coefficient. Figure 2.12b shows some critical signal path distances and plots critical wavelength as a function of system cutoff frequency.

2.3.2 TTL ICs in Industrial Environments

In an industrial environment the usefulness of noise margins at the system design level lies in the ability of a device to be impervious to noise spikes at the input. If an input voltage remains exclusively in the low-logic state, it can undergo any excursions within that state. A level change from 5.5 V to 2 V or from ground to 0.8 V should not affect the output state of the device.[27] For very noisy environments, the so-called *high threshold logic (HTL)* devices will be the answer. An example of such devices is the *Motorola high threshold logic (MHTL)*[28] that has noise margins of 5 V in both the high and low states for a V_{CC} of 15 V.

The ability of a logic element to operate in a noise environment involves more than the dc or ac noise margins. As device logic levels change, circuit currents flowing also change. These changes are due to different currents required to maintain the new logic level, transients due to lines charging and discharging and conduction overlap of the transistors in a TTL output stage.

Figure 2.13a shows the stray circuit elements associated with these transients. Capacitance C_s is the capacitance to ground and surroundings of the output connection from gate G1 to the input of gate G2. Lg and Ls are the self-inductance of the ground and supply line respectively. For standard TTL with typical rise and fall times of 6 ns or less, interconnection charging currents can be greater than 10 mA per output. This combined with overlap in the output stage indicates that care should be taken with system layout in order to minimize supply line impedance and interconnection inductance.

Noise that has entered the system, as well as that which is internally generated, can couple into adjacent signal and power lines. This can take place through common impedances as shown in Fig. 2.13b, due to mutual capacitance and inductance C_M and L_M.

The designer of TTL-based circuitry should be familiar with

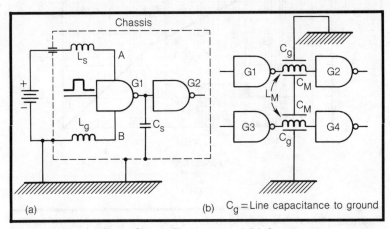

Figure 2.13—(a) Stray Circuit Elements and (b) Common Impedances in Connections of TTL Circuits

Logic Interfacing

emitted and conducted interference levels. Figure 2.14a shows the energy available for radiation from a typical TTL gate driving six loads (9.6 mA), at a 1 MHz frequency with a 50 percent duty cycle. Rise and fall times are assumed to be 6 ns. What is important in

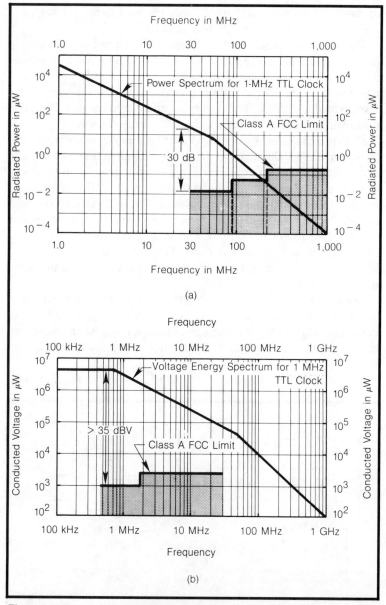

Figure 2.14—(a) Emitted and (b) Conducted Interference Levels for TTL Gate Compared to FCC Specification Limits. (From Ref. 29. Copyright by *Computer Design*, January, 1982. All rights reserved.)

the situation is this: Even if only 1×10^{-3} of the power available at the TTL gate was radiating energy into space, the radiated level would exceed FCC rules and regulations. Figure 2.14b shows an analogous problem for conducted interference. When power levels are much higher than those contained in TTL circuits, such as in a switching power supply, the situation becomes worse.[29]

Although each particular design case may need a special approach in fighting emitted and conducted interference, some general design rules for minimizing such interference in TTL systems are given below:

1. Power supply; ripple <5 percent and regulation <5 percent. RF filtering of primary supply.
2. Decouple every 2 to 4 packages with low-inductance ceramic disk capacitors of value 0.01 to 0.1 μF.
3. A ground plane is desirable when the printed circuit board contains a large number of packages. If no ground plane is used, the ground line should be as wide as possible and the area around the board periphery used as a distribution system.
4. Data rise and fall times into TTL devices should be <50 ns except in case of Schmitt triggers.
5. Unused inputs of all TTL devices should be tied to guaranteed logic levels, e.g., inputs of NAND/AND gates may be tied to V_{CC} where V_{CC} is guaranteed to *always* be <5.5 V. Otherwise, tie to V_{CC} through a series resistor; several inputs may share a common resistor.
6. Increased fan out can be obtained by paralleling devices in the same package.
7. Gate expanders should lie as close as possible to the gate being expanded. This avoids capacitive loading and noise pickup.
8. Input data to master slave JK flip-flops should not be changed when the clock is high.
9. If a ground plane is not used, the ground distribution should be arranged as an interlinking mesh.
10. Gates that drive back plane wiring via an edge connector should be mounted near the connector ground in order to provide a low impedance return for line currents.[27]

2.3.3 Wiring ECL Gates and GaAs Devices at the Printed Circuit Board Level[4,5,30]

Any signal path on a circuit board may be considered a form of transmission line. If the line propagation delay is short with respect to the rise time of the signal, any reflections are masked during the rise time. However, when the delay time in the wire is longer than the rise time of the input pulse, the reflected power causes an *overshoot* or *ringing* inside the line that affects the pulse as shown in Fig. 2.15a. Connecting the output resistor at the end of the connecting lead can help to reduce *ringing* (Fig. 2.15b).

Circuit designers may choose between transmission lines and conventional interconnect wiring when the distances between ECL devices are short, less than about 20 cm, or when the rise times are greater than 3 ns (e.g., MECL III devices). In many cases where ECL devices are used, transmission line techniques are advantageous. When using ECL devices with rise times less than 3 ns, transmission lines are highly recommended. The basic factors which will affect this decision are system rise time, interconnect

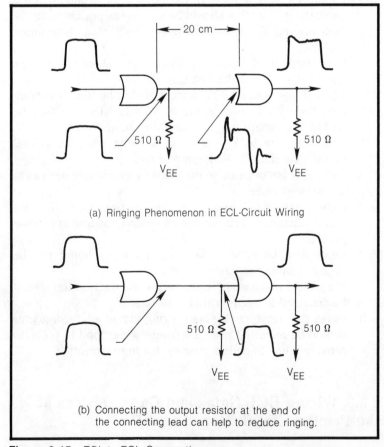

(a) Ringing Phenomenon in ECL-Circuit Wiring

(b) Connecting the output resistor at the end of the connecting lead can help to reduce ringing.

Figure 2.15—ECL-to-ECL Connections

distance, capacitive loading (fanout), resistive loading (line termination) and percentage of undershoot and overshoot permissible (i.e., reduction in ac noise immunity).

Some of the types of transmission lines that can be used for interconnecting high speed logic systems are shown in Fig. 2.16. A *twisted pair line* and the cross section of a *coaxial cable* transmission line are shown in Fig. 2.16a. Twisted pairs can be made from standard wires (AWG 24–28) twisted about 1 turn per centimeter. A twisted pair line presents a *characteristic impedance (Z_o)* between 100 and 120 Ω. Some popular types of coaxial cable have characteristic impedances of 50, 75, 93 or 125 Ω. Twisted pair and coaxial cables are suitable for long line lengths in the backplane, although twisted pair is susceptible to proximity effects and does not support data lines at as high a frequency.

Figure 2.16b depicts a cross section of a *wire over a ground* with characteristic impedance.

$$Z_o = \frac{60}{\sqrt{e_r}} \ln\left(\frac{4h}{d}\right) \tag{2.4}$$

where e_r is the *effective dielectric constant* surrounding the wire. This type of line is most useful for breadboard layout and for backplane wiring. The typical value of characteristic impedance is about 120 Ω which can vary as much as ±40 percent.

Another form of transmission line is the *microstrip line* (Fig. 2.16c) which is a strip conductor separated from ground plane by a dielectric. The characteristic impedance of a microstrip line that can be controlled to within ±5 percent is given by:

$$Z_o = \frac{87}{\sqrt{e_r + 1.41}} \ln\left(\frac{5.98\,h}{0.8\,w + t}\right) \; \Omega \tag{2.5}$$

The propagation delay (t_{pd}) of the line may be calculated by:

$$t_{pd} \approx 0.033 \sqrt{0.475 e_r + 0.67} \quad \text{ns/cm} \tag{2.6}$$

Finally, Fig. 2.16d shows a *stripline* which consists of a copper ribbon centered in a dielectric medium between two conducting planes. Its characteristic impedance that can also be controlled to within ±5 percent is calculated by:

$$Z = \frac{60}{\sqrt{e_r}} \ln\left(\frac{4b}{0.67\,\pi w\,(0.8 + \frac{t}{w})}\right) \; \Omega \tag{2.7}$$

Logic Interfacing

and is accurate enough when $W/(b - t) < 0.35$ and $t/b < 0.25$. The propagation delay of the stripline can be found using the equation:

$$t_{pd} \approx 0.033 \sqrt{e_r} \text{ ns/cm} \qquad (2.8)$$

From an implementation viewpoint, GaAs ICs present some difficulties when placed on PC boards. Board layouts must be more precise than standard PC boards, since line lengths, line separations, interconnections, impedances, terminations and

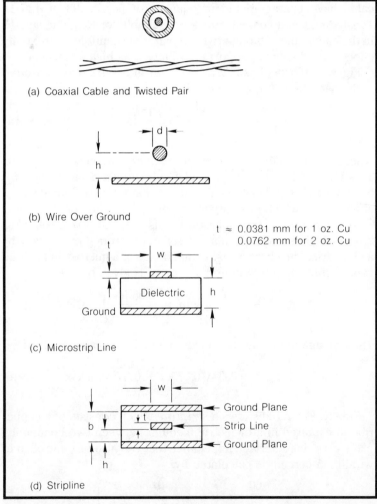

Figure 2.16—Various Types of Transmission Lines for PCB Connections

propagation delays become much more critical as clock rates move into thousands of megahertz. Solid dc and RF grounding of a GaAs IC package and RF "isolation" leads, as well as adequate heatsinking, are necessary to support performance in what is essentially a microstrip design medium.

2.3.4 CMOS Interconnections

The transfer characteristic of a typical CMOS logic gate is as shown in Fig. 2.17. The high input impedance of the gate results in no dc loading on the output so that the input and output signals are allowed to swing completely from zero volts (logic "0") to V_{DD} (logic "1"). These curves illustrate the high noise immunity of CMOS devices, i.e., typical 45 percent of the supply voltage.

The dissipation, P_{ac}, that results from the current components of a CMOS circuit is directly proportional to the frequency of operation (f), the amount of capacitive load (C) and the operating supply voltage (V). It may be expressed as follows:

$$P_{ac} = CV^2 f \quad (2.9)$$

In a CMOS device, all input terminals must be connected to a voltage level between the V_{SS} (the most negative potential) and V_{DD} (the most positive potential). In circuit configurations in which input terminals connect directly to NAND gates, the *unused inputs* must be tied to V_{DD} (high-level state) or tied together with another input terminal to a signal source. Conversely, in circuits which connect directly to NOR gates, the unused inputs must be tied to V_{SS} (low-level state, usually ground) or connected together to another input terminal driven by a signal source.

Certain usage rules must be observed for 54HC/74HC outputs as well as for inputs. Output voltages should not exceed the supply voltage and currents in the output diode should not exceed 20 mA. Moreover, output RMS drive currents should not exceed 25 mA for 4 mA standard-output devices or 35 mA for 6 mA devices.[17] In general, a risk of *latch-up* exists with CMOS devices due to the creation of a low-impedance path between the power-supply rails by the triggering of parasitic, four-layer bipolar structures (SCRs) inherent in CMOS input and output circuitry. Of course, latch-up will rarely be a problem in applications where the CMOS devices input and output pins never see voltages beyond the power-rail

Logic Interfacing

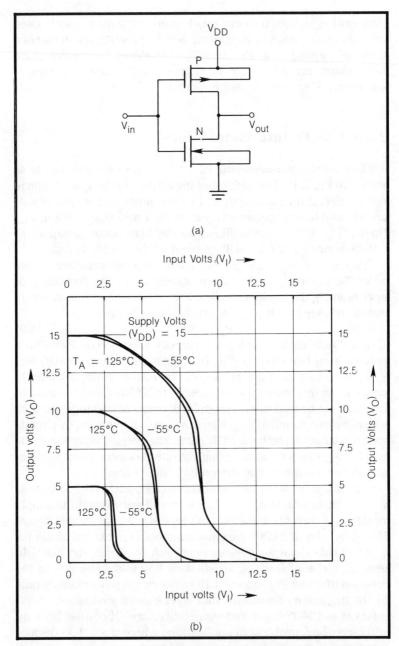

Figure 2.17—(a) Circuit Diagram of CMOS Inverter and (b) Voltage Transfer Characteristic of CMOS Inverter

values. The danger does exist, however, in systems exhibiting any combination of the following conditions:[31]

1. System operation/maintenance procedures allow insertion or removal of printed circuit boards with power applied.
2. The system is powered by multiple supply voltages (±12 V, +5 V and ground), or has multiple supplies of the same voltage (5 V regulated and unregulated).
3. Circuits use complex capacitive-decoupling techniques (particularly in the case of multiple supplies).
4. ICs on one system printed circuit board drive devices on other boards via a backplane or ribbon cable, for example.
5. Devices drive high capacitive leads such as long data or address buses.
6. The system contains high speed address or data buses that are long enough to make the effects of buses' inductive properties significant at the frequencies in question (ribbon cables are a prime example).
7. The system has electrical inputs that are directly accessible by the system's end user.
8. Analog devices, powered from higher supply voltages, drive the CMOS digital ICs.[31]

Plugging a circuit card into a live system with multiple power-supply voltages can result in overvoltage applications to some devices. Indenting the printed circuit board's edge connections can alleviate the problem. In Fig. 2.18,[31] for example, the topology shown ensures that the printed circuit board makes connection to ground first, then to +5 V, and finally to ±12 V supplies. The

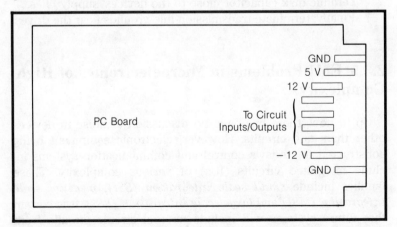

Figure 2.18—Indentation of Power-Supply Connections Ensures Proper Sequence of Pin Engagement (Reproduced here by Permission of Cahners Publishing Co., Division of Read Holdings, Inc., and Author P. Moennone)

Logic Interfacing

ascending-magnitude sequence assures that no over-voltages occur, even if one of the supplies pulls another through the decoupling capacitors.

As in previous cases, some general design guidelines are given here which are specially applicable to HC and HCT devices:[17]

1. Keep V_{CC}-bus routing short. When using double-sided or multi-layer circuit boards, use strip-line, transmission-line or ground-plane techniques.
2. Ground lines should be kept short, and the printed circuit boards they serve should be made as wide as possible even if trace width varies. Use separate ground traces to supply high-current devices such as relay and transmission-line drivers.
3. When local regulators are used, their inputs should be bypassed with a tantalum capacitor of at least 1 µF, and their outputs with a 10 to 50 µF tantalum or aluminum electolytic capacitor.
4. If the system uses a centralized regulated power supply, use a 10 to 20 µF tantalum electrolytic capacitor to decouple the V_{CC} bus connected to the circuit board.
5. Provide localized decoupling. For random logic, a rule of thumb dictates approximately 10 nF (spaced within 12 cm) per every two to five packages, and 100 nF for every 10 packages. These capacitances can be grouped, but it's more effective to distribute them among the ICs.
6. For circuits that drive transmission lines or large capacitive loads (microprocessor buses, for example), use a 10 nF ceramic disk capacitor close to the devices' supply lines.
7. Finally, terminate transmission-line grounds near the drives.

2.3.5 EMI Problems in Microelectronics of High Complexity

Up to now, there has been no discussion of noise in devices other than gate circuits. However, electronic equipment being constructed for today's control and communication systems include integrated circuits (ICs) of various complexity. These families include *small scale integration (SSI)*, *medium scale integration (MSI)* and *large scale integration (LSI)*. It is certain that future systems will include microelectronic circuits from families of greater complexity. Among these families will be *very large scale integration (VLSI)* and *very high speed integrated circuits (VHSIC)*. The trend toward increasing use of microelectronic circuits of greater complexity affects the requirement that electronic equipment operate properly in specified *electromagnetic (EM)* environments.[32]

However, the designers' EMC knowledge of microelectronic devices is limited to SSI and MSI. Information on the EMC properties of LSI, VLSI and VHSIC microelectronic circuits is essentially nonexistent. For present systems, a significant reduction in electromagnetic cross coupling can be obtained by the proper selection of IC devices used and their installation. Measurements indicate that geometrical factors, such as conductor distance above the ground plane, affect isolation measurements. Table 2.7 summarizes design recommendations as a result of testing.[33]

Table 2.7 EMI Design Recommendations for Microcircuits (Adapted from Ref. 33)

Design Parameters	Recommendation
Microcircuit Parts Selection • Packaging Type • Pinouts	• Flatpacks preferred over DIPs • Inputs and Outputs on Opposite Side of Package
Device Installation Method • Lead Routing	• Short leads routed close to ground plane
Circuit Board Layout • Multilayer Boards	• Input etch paths routed on separate layers and shielded from output etch paths
Intended Signal Spectrum • Waveform Shaping • Logic Level	• Rise times lengthened when possible • Low voltage logic selected when possible

2.3.6 Susceptibility of ICs to Electrostatic Discharge Damage

Electrostatic discharge (ESD) can cause permanent and temporary failure in microelectronic devices. Protection networks are used on electrostatic discharge sensitive devices in order to raise their immunity to damage from ESD.

Logic Interfacing

Both metal-oxide semiconductors (MOS) and bipolar discrete and monolithic device structures may be affected by ESD (see Table 2.5). In the case of bipolar devices, a representative scheme that provides transient protection is shown in Fig. 2.19.[23] It incorporates a transient-current sense resistor which, in conjunction with the diffused pocket-to-substrate junction capacitance, also serves to limit the leading edge pulse seen by active circuit.

Figure 2.20a shows a double diode ESD protection scheme for CMOS.[20] The network features the distributed diode input-resistor configuration (resistor value is 2 to 4 kΩ) and may employ an additional optional input diode, depending on manufacturer. It is shown by the dotted line, and is connected to V_{SS}. The improved B-series input/output ESD protection circuit of Fig. 2.20b is utilized in most RCA CD 4000B-series CMOS types.

Another CMOS protection circuit is shown in Fig. 2.20c. This circuit permits input signals in excess of V_{CC} to bring about down-level conversion, i.e., input logic levels of up to 18 V maximum may be converted to 3 V output logic levels. The input diodes to V_{DD} are eliminated to make this capability possible.

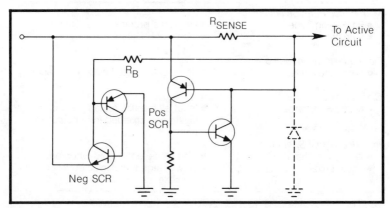

Figure 2.19—SCR Protection Structure Incorporating Transient Current Sense Resistor

ESD Damage

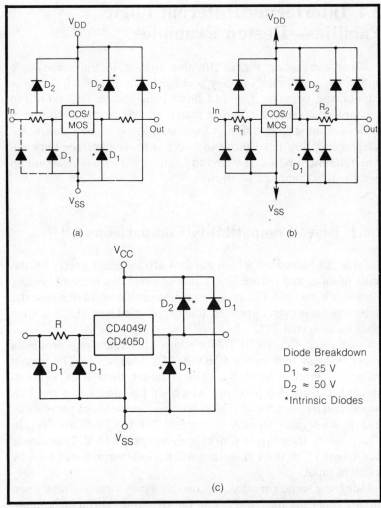

Figure 2.20—CMOS ESD Protection Circuits: (a) Standard-Gate-Oxide Protection Network, (b) Improved Gate-Oxide Protection Network, and (c) Protection Circuit Permitting Level-Shifting Function. (Reproduced by Permission of Electronic Engineering, © Morgan-Grampian Publishers, Ltd., and Author R. Funk)

2.4 Interfacing Different Logic Families—Design Examples

When connecting digital ICs that belong to the same logic family, there will rarely be any problems as long as such things as fan-out and parasitic line and input capacitance are taken into account. It is quite a different matter, however, to try to match devices coming from different logic families. In many cases, extra adapter circuitry is required to ensure proper interface between two different families. The circuitry can vary from a simple pull-up resistor to a complete IC device.

2.4.1 Brief Compatibility Comparisons [9,34,35]

Table 2.8 illustrates which families are matched purely on the basis of input and output levels. It is obvious that interconnecting elements within the TTL group cause no problems. In one case the noise margin is even improved; this happens if LS or ALS is used in place of standard TTL.

Connecting TTL to HCT MOS does not create any problems either because high speed CMOS is TTL compatible. The supply voltage tolerance with HCTMOS is larger than with TTL (10 percent instead of 5 percent), which simply means that the TTL supply can be used for HCTMOS but the reverse is not necessarily true. It is not quite so easy to connect TTL to CMOS. The V_{DH} in TTL is lower than V_{IH} in CMOS with a supply of 5 V. This means that a logic "1" at the TTL output will not be interpreted as *high* by a CMOS input.

Using the same practice, isolate the various parameters when mixing other families and define the required interfacing component or device needed for proper matching. Figure 2.21 is a summary of compatibility requirements of various logic families when interfacing one to another. The TTL family shown in Fig. 2.21 also includes Schottky and low-power Schottky versions.

In the following subsections some design examples of interfacing devices from different logic families are provided.

Table 2.8 Input and Output Characteristics of Various Logic Families (Adapted from Ref. 34)

	TTL	LSTTL	ALSTTL	CMOS	HCTMOS	HCMOS
V_{CC}	5V ± 5%	5V ± 5%	5V ± 5%	3...18V	5V ± 10%	2...6V
				5V		5V ± 10%
V_{IH}—logic 1 input level (min.)	2.0V	2.0V	2.0V	3.5V	2.0V	3.15V
V_{IL}—logic 0 input level (max.)	0.8V	0.8V	0.8V	1.5V	0.8V	1.1V
V_{OH}—logic 1 output level (min.)	2.4V	2.7V	2.7V	4.5V	3.7V	3.7V
V_{OL}—logic 0 output level (max.)	0.5V	0.5V	0.4V	0.4V	0.4V	0.4V
I_{IL}—input sink current (max.)	−1.6 mA	−0.36 mA	−0.2 mA	0.005 µA		
I_{IH}—output source current (max.)	40 µA	20 µA	20 µA	0.005 µA		
I_{OL}—output sink current (min.)	16 mA	8 mA	4 mA	0.4 mA	4 mA	4 mA
I_{OH}—output source current (min.)	−400 µA	−400 µA	−400 µA	−0.4 mA	−4 mA	−4 mA

Logic Interfacing

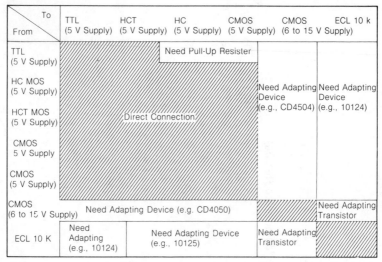

Figure 2.21—Compatibility Requirements of Various Logic Families

2.4.2 ECL/TTL Interfaces[36]

The most common interface requirement for ECL is with TTL logic levels. This occurs when an ECL system must interface with an existing TTL system, or when both ECL and TTL are used in the same system design. The interface requirements between ECL and TTL depend on how the circuits are being used. The normal MECL/TTL interface occurs when ECL is powered with −5.2 V power supply and TTL with +5 V. The use of common ground and separate power supplies helps isolate TTL generated noise from the ECL supply lines.

In many system designs where a small number of ECL circuits are used, it is desirable to operate both ECL and TTL on a +5 V power supply. ECL works very well in this mode if care is taken to isolate the TTL generated noise from the ECL +5 V supply line. In such circuits, translator circuits provide proper interfacing. The TTL-to-ECL translator shown in Fig. 2.22a consists of three resistors in series to attenuate TTL outputs to ECL input requirements. The translation is very fast, normally under 1 ns, depending on wiring delays and stray capacitance.

Two methods of interfacing ECL-to-TTL are illustrated in Figs. 2.22b and 2.22c. The circuit in Fig. 2.22b takes advantage of ECL complementary outputs to drive a differential amplifier made of

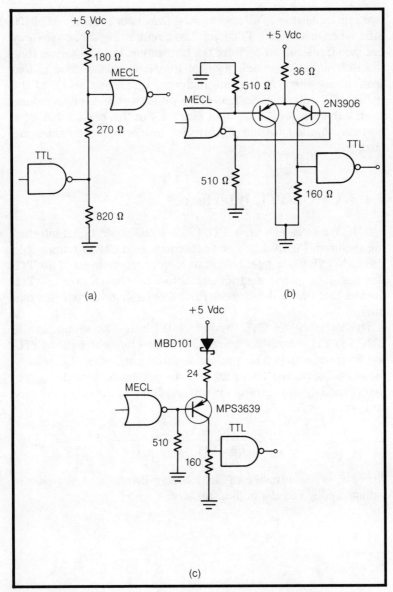

Figure 2.22—Common Supply ECL/TTL Interface Circuits

Logic Interfacing

two npn transistors. The speed of the translator is in excess of 100 MHz when driving one TTL load. The circuit in Fig. 2.22c uses only one pnp transistor to perform the translation, but it is slower than the differential approach. Typical translation delay time is less than 10 ns when driving one high-speed TTL load. Both of the ECL-to-TTL interface designs use a pull-down resistor to ground to sink the low level TTL input current. For this reason, fanout is normally limited to one TTL device unless resistor values are changed.

2.4.3 CMOS/TTL Interfaces[9]

CMOS devices can drive TTL loads with no additional interfacing required. Figure 2.23a is a schematic of a CMOS output gate driving a TTL input gate. The input current requirement of the TTL devices does place a strict limitation on the number of TTL devices that CMOS devices can drive from a single output (the fanout).

The interface for TTL driving CMOS is not as simple as the CMOS-to-TTL interface. Figure 2.23b shows the schematic of TTL-to-CMOS interface. The pull-up resistor eliminates the voltage incompatibility. The lower limit of the pull-up resistor, $R_{P\ min}$, is determined by the current-sinking capability of the driving device and is:

$$R_{P\ min} = \frac{V_{CC} - V_{OL\ max}\ (TTL)}{I_{OL}\ (TTL) + n\ I_{IL}\ (load)} \quad (2.10)$$

where n is the number of loads being driven, and V_{CC} is the voltage applied to the pull-up resistor.

CMOS/TTL

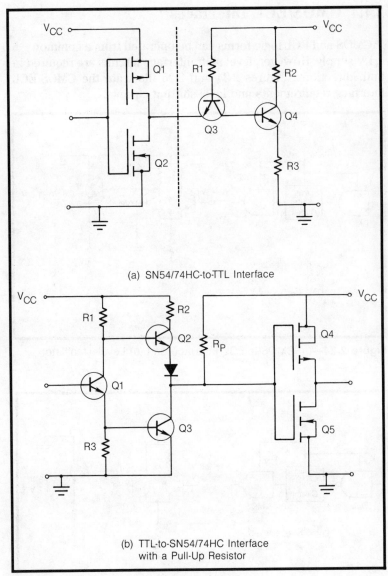

Figure 2.23—CMOS/TTL Interfaces

Logic Interfacing

2.4.4 CMOS/ECL Interfaces[37]

CMOS and ECL logic forms can be operated from a common -5 ± 1 V supply. However, level-shift interface circuits are required in both directions. Figures 2.24 and 2.25 illustrate the CMOS/ECL interface requirements and level-shifting circuits.

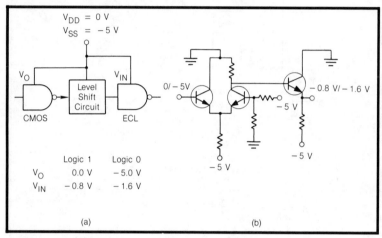

Figure 2.24—(a) CMOS-to-ECL Interface and (b) Level Translator

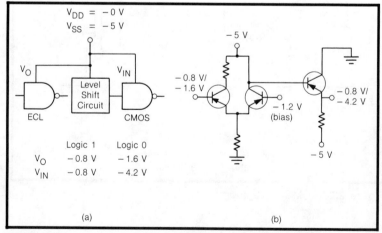

Figure 2.25—(a) ECL-to-CMOS Interface and (b) Level Translator

To interface with 10K ECL logic, the 10124 TTL-to-ECL and the 10125 ECL-to-TTL translator ICs can be used. It should be noted that these devices operate at TTL levels. When working with the 10125, a pull-up resistor must be used in accordance with the rules for driving HC devices from TTL sources. If an HCT device is used, the pull-up resistor after the translator IC is eliminated (see Fig. 2.21).

2.4.5 HSCMOS/CMOS Interfaces[7,35]

HSCMOS ICs can be connected directly to standard 4000B series CMOS ICs if they operate from the same supply voltage. Level-shifting may be required if the circuits are powered by different supplies. The 4504 or 4104 low-to-high level shifter, for example, can shift either TTL-output logic levels or CMOS logic levels to a higher output. The 4000B CMOS logic family operates up to 18 V. The HC and HCT families accept up to a 6 V supply and a minimum of 2 V. The 4000B series can be linked with HC and HCT circuits using the 4049, 4050B or HC4049 and HC4050 buffer ICs.

Of course, other equivalent level-shifting devices are available and can be used by carefully observing the power-supply connecting rules. Figure 2.26a shows a standard CMOS-to-HSCMOS interface with up/down voltage level conversion between any two power supplies. Figure 2.26b illustrates a standard CMOS-to-HSCMOS with high-voltage to low-voltage level conversion. The high-to-low translators HC4049 and HC4050 are useful for interfacing metal-oxide CMOS to HSCMOS devices because a voltage higher than the supply voltage can be applied to the inputs without risking damage to the device.

Logic Interfacing

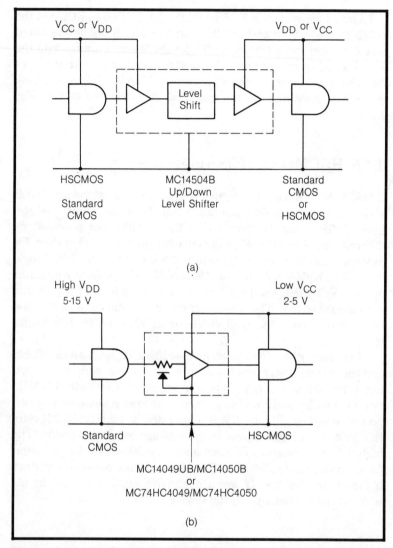

Figure 2.26—(a) Standard CMOS-HCMOS Interface with Up/Down Voltage Conversion and (b) Standard CMOS-HSCMOS Interface with High-to-Low Voltage Conversion

2.5 References

1. RCA Solid State Division. *Interfacing COS/CMOS With Other Families, Application Note ICAN-6602.*
2. R. A. Stehlin, *Two Schottky TTL Families, Computer Design*, July 1980, p. 154.
3. W. T. Greer, Jr. and B. Bailey, *Speed Up, Power Down for Bipolar Logic, Computer Design*, November 1984, p. 161.
4. Balph, T. *Implementing High Speed Logic on Printed Circuit Boards*, Proceedings of Wescon/81, San Francisco, CA, September 15–17, 1981.
5. W. R. Blood, *MECL System Design Handbook*, Motorola Semiconductor Products, 1980.
6. J. Fiebiger, *C-MOS: A Designer's Dream With The Best Yet to Come, Electronics*, Vol. 57, No. 7, pp. 113–115, April 5, 1984.
7. J. King, *Designing Tomorrow's Systems With High-Speed CMOS Logic, Proceedings of Midcon/83*, Chicago, IL, September 13–15, 1985.
8. T. R. Smith, *Advanced Logic Technologies Boos Mil-System Performance, Defense Electronics*, July 1984, p. 82.
9. Texas Instruments, *High-Speed CMOS Logic Data Book—Silicon-Gate Complementary MOS*, European Edition, 1984.
10. F. Wanlass and V. Arat, *Fastest-Yet CMOS Family Makes Its Bid For More Schottky Territory, Electronic Design*, February 7, 1985, p. 103.
11. B. Hoffman, *Finding A Home For GaAs ICs, Microwaves*, Vol. 24, No. 7, July 1985, pp. 43–46.
12. P. T. Greiling, R. E. Lundgren and C. F. Krumm, *Why Design Logic With GaAs—and How? MSN*, January 1980, p. 48.
13. S. Sando, *Using Nanosecond RAMs*, Proceedings of Electro/85, New York, NY, April 23–25, 1985.
14. A. Firstenberg and S. Roosild, *GaAs ICs for New Defense Systems Offer Speed and Radiation Hardness Benefits, Microwave Journal*, March 1985, p. 145.
15. R. C. Eden, A. R. Livingston and B. M. Welch, *Integrated Circuits: The Case for Gallium Arsenide, IEEE Spectrum*, December 1983, p. 31.
16. Texas Instruments, *Advanced Low-Power Schottky—Advanced Schottky*, Supplement to TTL Data Book, Vol. 2, 1984.
17. L. Wake, *High-Speed CMOS Designs Address Noise and I/O Levels, EDN*, April 19, 1984, p. 285.

18. R. Chao, *CMOS Logic Families Diversify Design Options*, Electronic Design, December 8, 1983, p. 203.
19. J. Binneboese, *Interfacing Enters A New Generation*, Proceedings of Southcon/82, Orlando, FL, March 23–25, 1985.
20. R. Funk, *Susceptibility of Semiconductors to Electrostatic Damage*, Electronic Engineering, March 1983, p. 51.
21. C. J. Georgopoulos and V. C. Georgopoulos, *Electrostatic Discharge—Its Hazardous Effects in Working Environments*, EMC Technology, July-September, 1984, p. 63.
22. J. K. Keller, *Protection of MOS Integrated Circuits From Destruction by Electrostatic Discharge*, Electrical Overstress/Electrostatic Discharge Symposium Proceedings, San Diego, CA, September 9–11, 1980.
23. L. R. Avery, *Electrostatic Discharge: Mechanisms Protection Techniques, and Effects on Integrated Circuit Reliability*, RCA Review, Vol. 45, June 1984, pp. 292–301.
24. C. J. Georgopoulos and V. C. Georgopoulos, *Electrostatic Control in Modern Electronic Office Communication Environments*, MECO '83 Int'l Symposium Proceedings, Aug. 29—Sept. 2, 1983.
25. J. DeFalco, *Reflections and Crosstalk in Logic Circuit Interconnections*, IEEE Spectrum, July 1970.
26. R. K. Southard, *Interconnection System Approaches for Minimizing Data Transmission Problems*, Computer Design, March 1981, pp. 107–116.
27. B. Norris, *Power-Transistor and TTL Integrated-Circuit Applications*, Texas Instruments Electronic Series, McGraw-Hill Book Company, N.Y., 1977.
28. Motorola, *Noise Immunity With Motorola High Threshold Logic*, Application Note AN-298.
29. E. VanderHeyden and J. H. Bogar, *Designing for Compliance With FCC EMI Regulations*, Computer Design, January 1982, p. 189.
30. T. J. Tyers, *ECL Logic Circuits*, Radio Electronics, Nov. 1983, p. 65.
31. P. Mannone, *Careful Design Methods Prevent CMOS Latch-Up*, EDN, January 26, 1984, p. 137.
32. S. S. Whalen, *Determining EMI in Microelectronics—An Overview*, 1981 IEEE International Electromagnetic Compatibility Symposium Record, Boulder, CO, Aug. 18–20, 1981.
33. K. M. Belisle and C. K. Jackson, *EMI Design Techniques for Decoupling and Isolation of Microcircuits*, 1983 IEEE International Electromagnetic Compatibility Symposium Record, Washington, D.C., August 23–25, 1983.

34. *Mating Logic Families*, *Elektor*, February 1984, pp. 2.48–2.251.
35. R. E. Funk and J. P. Exallo, *Linking HC and HCT Chips With Other Circuits is a Question of Logic*, Electronic Design, June 6, 1985, p. 157.
36. B. Blood, *Interfacing With MECL 10,000 Integrated Circuits*, Application Note AN-720, Motorola Semiconductor Products, Inc.
37. RCA, *COS/MOS Integrated Circuits Manual*, NJ, 1972.
38. N. Edmundson, *ECL in High-Performance Systems*, Proceedings of Electro/85, New York, April 23–25, 1985.

General References

- S. A. DeFalco, *Comparison and Uses of TTL Circuits*, Computer Design, February 1972, pp. 63–68.
- T. Hiniker, *Noise Immunity of Capacitive Switch Improves With Flip-Flop Interface*, Electronic Design 15, July 19, 1978, p. 98.
- R. Pease, *Bounding, Clamping Techniques Improve Circuit Performance*, EDN, November 10, 1983, p. 277.
- L. Wakeman and K. Karakotsios, *New 54HC/74HC Silicon Gate CMOS Logic Family Matches Low Power Schottky Performance*, Northcon/82, Seattle, WA, May 18–20, 1985.
- V. Boan, *Designing Logic Circuits for High Noise Immunity*, IEEE Spectrum, January 1973, p. 53.
- B. Ay and S. C. Crist, *Design and Evaluation of a Bipolar Tri-State Logic Family*, IEEE Journal of Solid-State Circuits, Vol. SC-17, No. 1, February 1982, pp. 16–19.
- S. Caniggia, *EMC Design of Digital Systems Using Macromodelling Procedures for Integrated Circuits and Their Interconnections*, EMC 5th Symposium and Technical Exhibition on Electomagnetic Compatibility, Zurich, March 8–10, 1983.
- G. Schonk, *Quality—High-Speed CMOS*, Electronic Components and Applications, Vol. 7, No. 1, 1985, pp. 7–16.
- C. S. Hinkle and S. A. West, *A Noise Consideration in Fast Logic Lead Inductance*, Electro/84, Boston, MA, May 15–17, 1984.
- R. Q. Sproul, *Good RF Design Techniques Aid High-Speed TTL Control*, Microwaves & RF, November 1983, p. 111.

- P. Emerland, *Knowing Inductive Logic Keeps Power Interface ICs Alive and Switching*, Electronic Design, March 14, 1985 p. 195.
- D. MacMillan and T. Gheewala, *Learn Gallium-Arsenide Basics Before Applying High-Speed ICs*, EDN, March 22, 1984, pp. 239–244.
- H. W. Ott, *Digital Grounding and Interconnection*, 1981 IEEE International Electromagnetic Compatibility Symposium Record, Boulder, CO, August 18–20, 1981.
- W. D. Greason and G. S. P. Castle, *The Effects of Electrostatic Discharge on Microelectronic Devices—A Review*, IEEE-IAS Annual Meeting, Philadelphia, PA, October, 5–9, 1981.
- M. Martinez, *CMOS Logic With Bipolar-Enhanced I/O Rivals Fast TTL Gates*, Electronic Design, February 7, 1985, p. 117.
- B. Furlow, *Fewer Connections Translate Into Fewer Failures*, Computer Design, April 1985, p. 103.

Chapter 3
Line Drivers/Receivers and Interfacing Cables

The transmission of error-free data over long lines in noisy environments requires the use of special transmission-line drivers and receivers and the careful selection of a suitable transmission line.

3.1 Introduction

The requirement for using special transmission-line drivers and receivers is especially critical when data are transmitted between consoles, or between computer and remotely located peripherals.

Any design engineer faced with reducing radiated EMI must concentrate on preventing currents from flowing into the wires and ground shields of a system's cable connecting line drivers and receivers. It must be known, also, that the need to determine pulse reflections of interconnections and terminations—e.g., sockets, edge connectors and drive input/output ports—is just as important as supplying adequate pulse bandwidth.

The need for data transmission standards has become apparent as industry has matured. The U.S. Electronic Industries Association (EIA) has developed several specifications to standardize the interface in data communication systems. Among them are the RS-232, RS-423, RS-422 and RS-485 which provide the specifications for interconnecting single-ended and differential line drivers and line receivers in a system. Key considerations in selecting a data transmission standard are: line length (the distance between component elements), bit rate (the speed at which the data is to be transmitted), environment (noise conditioning along the transmission path) and whether or not the system will have to interface with other existing or future systems.

3.2 Signal Transmission Lines

The purpose of an interconnection line in any digital system is to transmit information from one point of the system to another. When information on a signal line changes, a finite amount of time is necessary for the information to travel from the sending end to the receiving end of the line. As the circuit speed becomes faster and clock rates increase, the dynamic behavior of the interconnection line becomes important. The rise and fall times of the logic elements, loading effects, delay times of the signal paths and various other transient characteristics affect reliable operation of the system.

3.2.1 Definitions of Line Transmission Factors

The following are some descriptive definitions concerning transmission line factors[1] which will help in understanding the subjects of subsequent subsections:

1. Characteristic impedance (Z_o) is a function of the geometry and the materials of the system. In a transmission line, two wires are surrounded by a dielectric material. Between the two wires is a capacitance (C_o), which is a function of the spacing between the wires, the dielectric constant (e_r) of the material between them and the effective plate area of the wires. In simple instances, plate area is equal to the diameter times the length of the wire. Capacitance can be expressed as microfarads (μF) per unit length. The two wires also have a *mutual inductance* (L_o) which is a function of the distance between the wires, their diameter and the wire length (the magnetic permeability of the surrounding material is usually 1). This inductance can be expressed as microhenries per unit length. In terms of the above elements, the characteristic impedance would be $Z_o = \sqrt{L_o/C_o}$ Ω.

2. The reflection coefficient (p) is a measure of how well the transmission line has achieved impedance control. A well-designed system in which the circuits at both ends of a line are matched to the impedance of the line itself, provides maximum energy transfer between the circuits. More importantly, for digital applications, a well-matched system prevents the generation of noise on the line, manifested as reflections at points where impedance changes. The amount of voltage reflected determines the voltage reflection coefficient, p. Reflections, which occur at impedance mismatches, are delayed by the propagation time of the cable, and may cause undesired circuit triggering.
3. Crosstalk is the amount of signal pickup on one inactive line from adjacent, driven lines in the same cable (for multiconductor cables), or from driven lines in adjacent cables. Such pickup signals are normally termed line-to-line and cable-to-cable crosstalk, respectively. In addition to representing a loss of signal from the driven circuit, crosstalk contaminates the adjacent, quiescent lines and may cause false triggering, overloaded circuits and interference.
4. Attenuation of digital pulses results in degradation or distortion of the pulse in terms of a loss in peak voltage, a slower pulse rise or both. There are two possible sources of these problems: *conduction loss* in the dielectric and *resistive loss* in the conductor. Both are functions of frequency; dielectric attenuation varies directly, whereas conductor loss (*skin effect*) varies as the square root of frequency. The sum of these two loss parameters represents the total loss of a line section and is usually measured in dB per unit length.
5. Propagation delay (t_{pd}) is the time required for a pulse to travel through a transmission line system. It is the reciprocal of signal propagation velocity and is, essentially, a function of the dielectric constant of the insulating material.

3.2.2 Digital Signal Line Models[2,3]

A digital signal line, using coaxial cable, can be modeled as a transmission line (Fig. 3.1),
where,

Drivers/Receivers

$$Z_o = \sqrt{\frac{L_o}{C_o}} \ (\Omega) \qquad (3.1)$$

$$t_{pd} = \sqrt{L_o C_o} = Z_o C_o = 0.033 \sqrt{e_r} \ (ns/cm) \qquad (3.2)$$

$$P_L = \frac{R'_L - Z_o}{R'_L + Z_o} \qquad (3.3)$$

R'_L is equivalent to $R_L // R_{in}$, where R_{in} is the input resistance of receiving gate B while the source V_s, in series with the resistance R_s, represents the output of transmitting gate A, in Fig. 3.1b.

Figure 3.2a shows a typical differential driver/receiver digital interface using as interconnection medium a twisted pair line. The line receiver and driver circuits and their interconnecting cable may be reduced into the elements of Fig. 3.2b.

The various interfacing elements in the equivalent circuit of Fig. 3.2a are defined as follows:
1. Z_{oga} and Z_{ogb} are impedance to ground at driver circuit output terminals. These elements include, also, the effect of near-end terminating resistor R_1.
2. $Z_{oda} + Z_{odb}$ is the differential output impedance of driver.

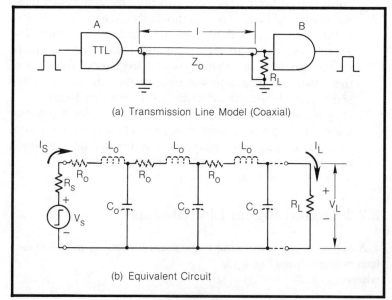

Figure 3.1—A Digital Signal Line with Coaxial Cable

Digital Line Models

3. Z_{tga} and Z_{tgb} are impedance to ground of twisted pair wires.
4. Z_{iga} and Z_{igb} are impedance to ground at receiver circuit input terminals.
5. Z_{id} is the differential input impedance of receiver circuit. This element includes the effect of far-end terminating resistor R_2.

Twisted pair lines differentially driven into a line receiver provide maximum noise immunity. Any noise coupled into a twisted pair line appears equally on both wires (common-mode). Because the receiver senses only the differential voltage between the lines, crosstalk noise has no detrimental effect on the signal up to the common-mode rejection limit of the receiver.

When using fast logic elements that may operate at clock rates that reach 150 to 200 MHz, what must be considered is how the 1 to 1 ns pulse edges propagate throughout the hardware, 1 to 1 ns especially through the interconnecting transmission line. Figure

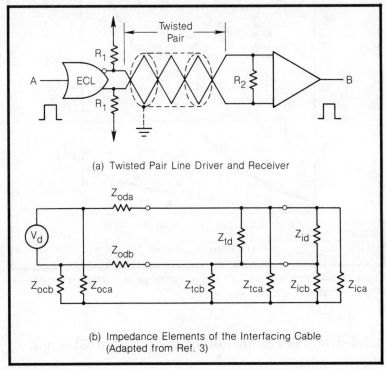

(a) Twisted Pair Line Driver and Receiver

(b) Impedance Elements of the Interfacing Cable
(Adapted from Ref. 3)

Figure 3.2—A Digital Signal Line with Twisted Pair Line

Drivers/Receivers

3.3 shows the signal edge rise time vs. distance in two different conductive media. It is clear that a 2 ns step will propagate up to about 1.1 m along a twisted pair without degradation, while the same signal can propagate to more than 3 m in coaxial cable without degradation.[4]

Other basic types of interconnections mainly found on printed circuit boards that can be modeled via transmission line techniques include: *wire over ground, microstrip* and *stripline*. These types were discussed in Chapter 2 of this book.

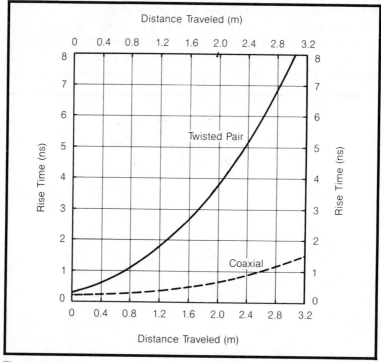

Figure 3.3—Signal Edge Rise Time vs. Distance in Two Interconnection Media (Adapted from Ref. 4)

3.2.3 Line Termination with Resistive Element

When the interconnections used to transfer digital information become long enough, so that line propagation delay is equal to or greater than the pulse transition times, line terminations must be considered.

Line Termination

The voltage across an impedance terminating a transmission line is a function of the real and imaginary components of the impedance, the characteristic impedance of the line and the incident power. When the impedance is a pure resistance—an assumption that simplifies this discussion—and the transmission line is ideal, then using Fig. 3.4a we can write:

$$P_R = P_I \left(\frac{R_L - Z_o}{R_L + Z_o}\right)^2, \tag{3.4}$$

$$P_L = P_I - P_R = P_I \left[1 - \left(\frac{R_L - Z_o}{R_L + Z_o}\right)^2\right], \tag{3.5}$$

$$V_L = \sqrt{P_L R_L} = \sqrt{I_L^2 R_L^2} = I_L R_L, \tag{3.6}$$

where, P_I = incident power $\quad P_L$ = power delivered to R_L
P_R = reflected power $\quad Z_o$ = line characteristic impedance
R_L = load resistance

If $R_L = Z_o$, the numerators of the fractional terms in Eqs. (3.4) and (3.5) become zero and the reflected power is zero. With reflections reduced to zero, one source of signal distortion and noise is eliminated. Equation (3.6) shows the relationship for P_L, R_L, V_L and I_L.

In line circuit design, R_L is a lumped value representing the combination of a termination resistor (R_T) and the input resistance (R_{in}) of the receiver, as shown in Fig. 3.4b. The voltage, V_L, can be expressed as follows:

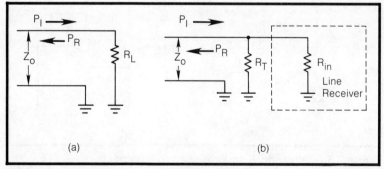

Figure 3.4—(a) Impedance Termination in Transmission Line and (b) The Parallel Combination of Termination Resistor (R_T) and the Input Resistance (R_{in}) of the Line Receiver Makes Resistor R_L

Drivers/Receivers

$$V_L = \sqrt{P_L \frac{R_{in} \times R_T}{R_{in} + R_T}} \qquad (3.7)$$

If $R_{in} \gg R_T$, then the incoming signal power and noise power are shunted to ground by R_T, decreasing the effective power to the input of the receiver.

3.2.4 Line Termination with Clamping Diodes[5,6]

In a digital-logic system, diode terminated interconnections are sometimes a better choice than resistively terminated lines. The diodes suppress multiple line voltage reflections almost as effectively as a resistor. Yet, unlike a resistor, they dissipate very little power and run no risk of overloading the logic circuitry. A typical logic interconnection circuit is shown in Fig. 3.5. This circuit configuration consists of a push-pull driver, a transmission line and a receiver/current switch. The input impedance of the switch is much greater than the characteristic impedance Z_o of the line, so in the absence of a termination element the line would see an open circuit. Under this condition, reflections on the line would upset switching performance and might damage transistors in both drive and receiver. By using terminating diodes (dotted lines in Fig. 3.5), the reflections are suppressed.

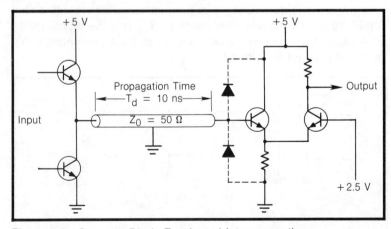

Figure 3.5—Common Diode Terminated Interconnection

Interfacing Problems

3.2.4 Line Termination with Clamping Diodes

Another termination scheme used in ECL systems uses Schottky diodes to clamp the voltage at the remote end of the line and suppress reflections (Fig. 3.6). The cut-in voltage of the diodes is approximately half an ECL logic swing. Since the diodes are returned to the threshold voltage V_{BB}, the quiescent power drain is small; however, any overshoot causes the diode slope resistance to decrease rapidly and, hence, the energy is absorbed.

The great benefit of reflection clamping with Schottky diodes is that it allows clean waveforms to be received over uncontrolled impedance lines. It is particularly useful when a few very high speed ECL III functions must be included in an ECL 10K design.

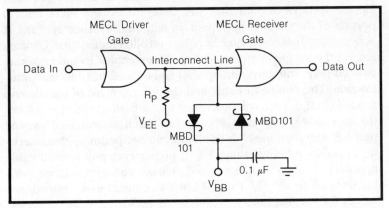

Figure 3.6—Schottky Diode Termination of a Line

3.3 Problems with Interfacing Cables and Connectors

Many new insulation materials and new cable configurations have been introduced to the industry; thus providing an opportunity for engineers to optimize packaging design for minimum size and weight. For the engineer who must match the cable to the specific application,[7] as is the case of line drivers and line receivers, an understanding of the basic characteristics of the various types available is essential. Poorly selected and installed cabling can act as both noise transmitting and receiving antenna or as undesired primary and secondary windings of coupling transformers, placing interference where it should not be.

Drivers/Receivers

On the other hand, in an interconnecting subsystem composed of cable connector and backshell accessories, the *connector* is the most susceptible to the electromagnetic interference. Here, again, designers must be familiar with applied methods, such as filtering and shielding, in order to circumvent the interference problem.

3.3.1 Basic Types of Cable Lines and Their Characteristics

Cables carry data between widely spaced sections of the subsystem where line drivers and line receivers are the end electronic devices, and they are likely to be the most critical portion of the transmission path. In high performance systems, it is necessary that the cables have a controlled impedance. Choices among the various controlled impedance cables include coaxial, twisted pair, standard and special ribbon and flexible flat transmission. The choice of cable and the arrangement of signals and grounds in the cable will determine transmission path characteristic impedance. Low cost approaches, such as standard twisted pair and standard unshielded ribbon, do not permit a characteristic impedance below about 80 Ω; higher cost and special cable approaches allow well-controlled characteristic impedance ranging from 50 to 150 Ω.[1] Table 3.1 shows characteristic impedance values for various lines.[4]

Several other cable characteristics are also important, depending upon the application. For instance, attenuation (signal loss per unit length, which causes rise time degradation), capacitance per unit length, pulse signal propagation delay time and crosstalk must be considered.

Standard ribbon cable construction permits a good deal of control over transmission path spacing dimensions and signal length in parallel data systems. For better results, standard cables

Table 3.1 Characteristic Impedances for Various Lines

Type of Line	Value of Z_o
Coaxial cable	50 to 75 Ω
Twisted pair	100 to 120 Ω
Two lines, 2,5 cm apart	200 to 300 Ω
Circuit board tracks	50 to 150 Ω
Free Space	376 Ω

are driven with alternating signal and return lines for single-ended and differential drivers, as shown in Fig. 3.7. Crosstalk is one of the major limitations of ribbon cables. Even when carrying relatively slow rise time pulses, such as 10 ns for 6 m distances, the crosstalk can be higher than 30 percent for standard cables. Standard ribbon cables can be useful for short distances.[1]

Flat ribbon cables with dual dielectrics can significantly reduce the flat cable crosstalk.[8] Special ribbon and flat, flexible cables can meet requirements for transmission characteristics in high performance systems.

Twisted pair cables, unless they have close tolerance for both the wire and the twisting, have the disadvantage of a lower control

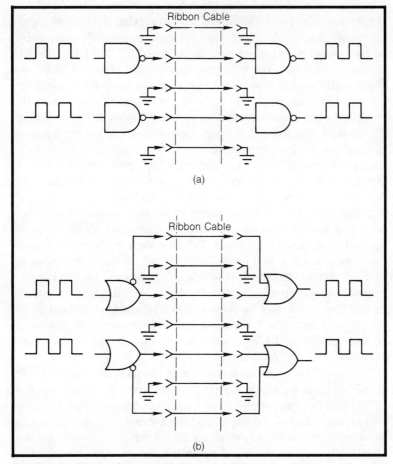

Figure 3.7—Ribbon Cable Applications: (a) Alternating Signal and Ground Return Lines and (b) Differential Drive

on their characteristic impedance and actual path length variations which introduces the possibility of pulse skewing. Characteristic impedance for twisted pair cables is about 100 to 120 Ω. These cables offer excellent shielding characteristics from low-frequency external fields, but are little better than cabled wire pairs in their susceptibility to crosstalk. Although they are acceptable for single-signal applications, they are not recommended for parallel data transmission.

In all cases of potential interference, low- or high-frequency, shielded cables should be used to protect against magnetic and capacitive stray fields. Such cables can be divided into four basic classes: coax, twinax, triax and quadrax cables.[9]

A *coax cable* consists of an inner and outer conductor, insulated from each other, with both conductors carrying the desired signal currents (source to load and return). Grounded coax cable installations are excellent and can be used for 20 kHz to 50 GHz for most systems. Low-frequency signals (20 kHz to 6 MHz) are particularly susceptible to both ground and common mode interference. In this case, coax cable is recommended with the complete coax chain having a *minimum* number of outer conductor ground contacts. Reducing the number of outer conductor ground connections reduces the number of possible ground loops. This demands that major equipment, relays, switches, connectors, patch panels, etc., be isolated from ground with the ultimate being one ground connection at the source (Fig. 3.8).

Twinax cable is a two-conductor twisted balanced wire line having a *specific impedance*, with a grounded shielding braid around both wires (Fig. 3.9a). Twisting the two balanced signal-carrying wires provides cancellation of any random induced noise voltage pickup, thereby giving protection against magnetic noise field, of the low-frequency variety, that passes through the copper braid. This cable also provides protection against ground loops and capacitive fields. Twinax cable usefulness, however, is limited to approximately 10 MHz since it has rather high transmission losses above this frequency.

Triax cable is coax cable with an additional outer copper braid, insulated from the signal carrying conductors, which acts as a true shield and protects the enclosed coax conductors. This braid, or shield, is grounded and bypasses both ground loop and capacitive field noise currents away from the signal carrying coax (Fig. 3.9b), thereby, greatly improving the *signal-to-noise* ratio over standard coax cable use. Triax cable is also used in *driven shield* applications where the inner conductor and first shield are driven in

Cable Types

parallel at the transmitting end, and work against the outer braid which is insulated above ground (Fig. 3.9c). At the receiving end, the inner braid is left floating, providing a *Faraday shield* between the inner conductor and outer braid. In this way the cable distributed capacitance is greatly reduced, thereby reducing cable losses and loading.

Quadrax cable is a twinax cable with two separate and insulated braids. The two braids are connected to *system* ground and

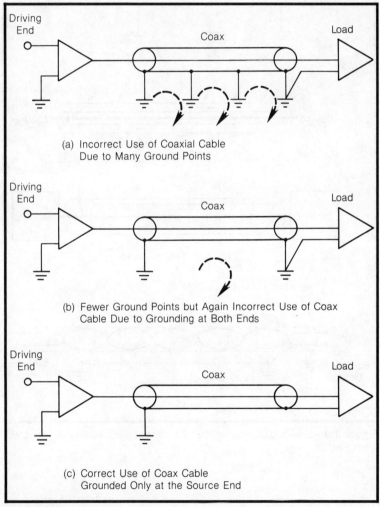

Figure 3.8—Grounded Coax Cables

Drivers/Receivers

earth ground, respectively, forming a *quadrax-guarded circuit* (Fig. 3.9d). Quadrax cable can also be used to provide additional noise and EMI suppression by connecting both shielding braids to earth ground if a separate equipment ground is not available.

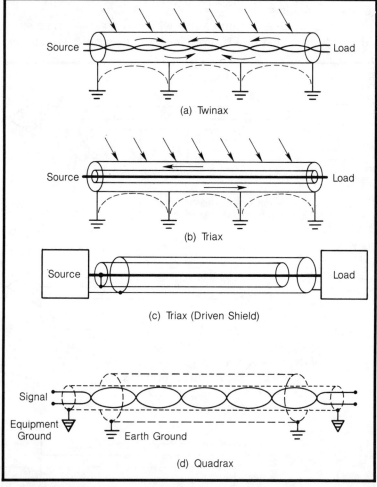

Figure 3.9—Shielded Cables for High-Interference Environments (Adapted from Ref. 9)

3.3.2 Figure of Merit for Shielded Twisted Pair and Coaxial Cables[10]

For a metallic transmission line (coax cable, twisted pair or twin-axial), a figure of merit, Fm_a, factor can be calculated provided that a measured attenuation coefficient at a known frequency is available. The only restriction is that f must be sufficiently high so that *skin effect* impedance dominates the loss (f ≥ 100 kHz for copper transmission lines).

A convenient expression for Fm_a is the following:

$$Fm_a = Rl^2 = 109 \frac{f}{a^2(\omega)} \tag{3.8}$$

where, R = megabits per second,
l = length in kilometers,
f = frequency in megahertz at which $a(\omega)$ is measured,
$a(\omega)$ = attenuation coefficient in dB/km.

The significance of this Fm_a is that it permits the comparison of the digital performance of different cables within a general cable class such as coaxial, twin-axial or twisted wire pair. For any given application requirement, a required Rl^2 product is easily calculated from the maximum desired bit rate and required distance. The Rl^2 product must be equal to or less than the Fm_a determined from Eq. (3.8) for a given transmission medium to be satisfactory.

In the cable configurations discussed so far, electrical pulses transmit the signal information—a form of transmission medium that has just about reached its limit of development. It is also questionable whether any further significant improvements are possible in crosstalk reduction, external-interference limitation and cable miniaturization. Optical transmission offers an alternative to these electrical cable types and can solve a lot of problems associated with conventional cable interconnections, as Chapter 8 will show.

Drivers/Receivers

3.3.3 Ground Loops

Ground loops are created when a chassis, frame or bus is used as a common ground for two or more pieces of equipment. Because no two ground points have exactly the same potential, more than one ground on a signal circuit or signal-cable shield creates a current flow between grounds (see also Fig. 3.10).[11]

The fields of any large *current loop* can be represented as a superposition of the fields from a number of smaller current loops having the same perimeter as the large loop. Thus, to effectively control radiated EMI from digital systems, one must effectively reduce the areas bounded by the current loops in such systems (Fig. 3.11).[11] For a quantitative understanding of the importance of minimizing current loop areas, consider the following equation:[12,13]

$$A = 380 \frac{Ed}{If^2} \qquad (3.9)$$

where A represents the current loop area in cm^2, E represents electric field strength in microvolts/meter, d represents distance

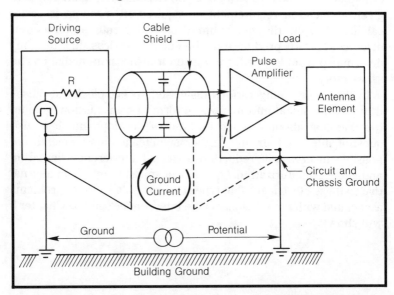

Figure 3.10—Ground Loops Formed by Interconnecting Electronic Equipment Located in Different Buildings

from the loop in meters, f represents the frequency of a current component in megahertz and I represents the amplitude of a current component in milliamperes. This equation has been derived from the expression for the maximum electric field of a small current loop, assuming the loop is positioned over a large conductive ground plane. It provides a good worst-case approximation of the electric field that would be observed from such a loop when measured according to FCC specifications.[14]

As an example, consider the loop areas which are associated with the FCC Class A and Class B radiated EMI limits for digital devices at 200 MHz. At that frequency the Class A limit is an electric field strength of 50 μV/m measured over a ground plane at a distance of 30 m; the class B limit is an electric field strength of 150 μV/m measured over a ground plane at a distance of 3 m.[14,15]

From Eq. 3.9, the loop area associated with the Class A limit for 1 mA current component at 200 MHz is 14.25 cm^2 and the loop area associated with the Class B limit, for the same current and frequency, is 4.275 cm^2. This implies that current loop areas must be minimized by digital designers if the FCC regulations are to be met.

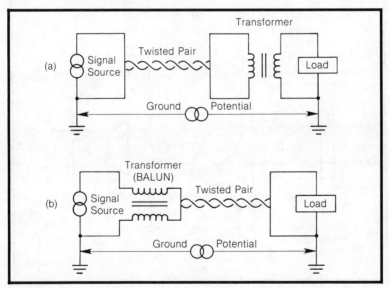

Figure 3.11—Elimination of Ground Loops

Drivers/Receivers

3.3.4 Crosstalk and Common-Mode Rejection

Coaxial cables present good common-mode (ground noise) rejection. The twisted pair, unless it is shielded, is not good because adjacent objects can distort the propagating signal. Common-mode rejection in twisted pairs is greatly improved if the line has balanced drive. To further minimize crosstalk effects, a differential amplifier can be used to select, compare and amplify low-level signals in noisy environments. In a differential mode, unlike signals applied to the double-ended input result in an output proportional to their difference (Fig. 3.12a). In the common-mode, like signals result in a negligible output. An interesting application is the use of differential amplifiers as line drivers and line receivers at the two ends of a transmission line (Fig. 3.12b).

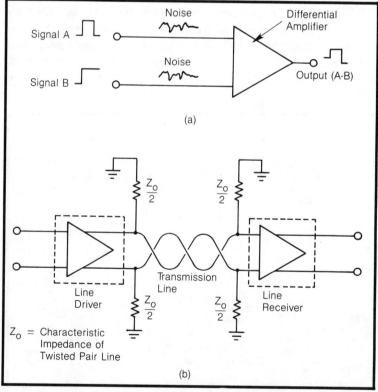

Figure 3.12—(a) Differential amplifiers reduce crosstalk by transmitting the useful signal while cancelling noise common to both inputs. (b) Amplifiers are often used at both ends of a transmission line in a driver/receiver configuration.

Crosstalk is significantly reduced because the twisted pair causes capacitive and inductive crosstalk to appear equally on both lines (common-mode), hence, enabling the line receiver to reject the undesired signals.

3.3.5 EMI/RFI Control at the Connector Interface

The connector can be visualized as a region in the cable shield containing various sizes and shapes of apertures through which radiated or conducted electromagnetic fields may penetrate. The basic circular design is two telescoping cylinders. The initial task of the designer will be to make these cylinders fit as closely as possible in mass production and still maintain reasonable mechanical tolerance. Under these conditions a continuous metal-to-metal contact around the circumference is extremely difficult to be implemented.[16]

When an electromagnetic field impinges on the surface of the cylinders, a current is driven from one cylinder to the other. This current will constrict at the points of contact causing an undesired voltage drop proportional to the specific contact resistances. It is common practice for connector designers to minimize this problem by using uniformly designed contact fingers properly joined to one of the mating cylinders. These fingers are designed to replace the partial contact between cylinders with uniform 360° contact, broken only by the necessary gaps between fingers.

Another method that aids reducing radiated EMI from cables is filtering.[17] *Capacitive filter connectors* are emerging as a reliable and effective technique. Not only can they suppress emitted RF noise currents, they can also minimize a system's susceptibility to externally induced noise currents. They restrict those spurious high-frequency currents at the connector—before they can leave or enter the equipment's enclosure. Electromagnetic *interference filters* are divided into three broad categories: those comprising *lumped elements*, those acting as special *feed through capacitors*, and those employing high-frequency, *lossy ferrites*.

Figure 3.13 illustrates the straight *feedthrough filter*. It depends on the capacitor to bypass high-frequency energy to ground, and to isolate input and output by the enclosure wall. Its equivalent circuit contains a series inductor that limits this effect at high frequencies.[18]

Drivers/Receivers

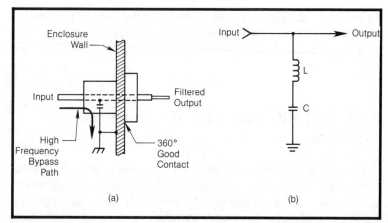

Figure 3.13—Connector Feedthrough Capacitor. Effective bypassing of RF currents depends mainly on capacitor (a), but parasitic inductor in equivalent circuit limits bypassing at high frequencies (b).

3.4 Characteristics and Applications of Line Drivers/Receivers

A *line driver* commonly translates the input logic levels (TTL, MOS, CMOS, etc.) to a signal more suitable for driving a transmission line. However, there are some logic families where gates may be used to drive the line directly (e.g., MECL).

A *line receiver* provides the reverse function of a line driver where the voltage that has been applied to a line is detected and restored to an output logic level. This function can be performed by various digital and linear ICs.

There are two basic means of communicating between components of a data processing system. These are *single-ended*, which use only one signal line for data transmission, and *differential*, which use two signal lines. Single-end transmission is used only for short distances and slower data rates since, as line length increases it is difficult to distinguish between a valid data signal and those signals introduced by external environment, such as ground shifts and noise signals. Differential data transmission overcomes these problems.

3.4.1 Description and Operation of Typical Line Drivers and Line Receivers[19]

Figure 3.14 is the schematic diagram of a typical line driver. The important features of this circuit are: (1) its outputs are current limited to protect the driver from accidental shorts in the transmission lines, (2) diodes at input and output pins clamp severe voltage transients which may be induced into the transmission lines and (3) the circuit has internal inversion to produce a differential output signal, reducing the skew between the outputs and making the output state independent of loading.

As can be seen from the upper half of Fig. 3.14, a quadruple-emitter input transistor, Q9, provides four logic inputs to the

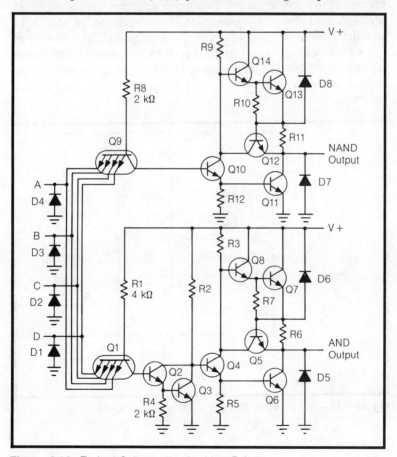

Figure 3.14—Typical Schematic of a Line Driver

Drivers/Receivers

driver. This transistor drives the inverter stage, formed by Q10 and Q11, to give a NAND output. In the high state, the output level is approximately two diode drops below the positive supply, or roughly 3.6 V at 25°C with a 5 V supply.

The lower half of Fig. 3.14 is identical to the upper, except that an inverter stage has been added, thus providing an AND output. This is needed so that an output signal which drives the output of the upper half positive will drive the lower half negative, and vice versa, producing a differential output signal.

A simplified schematic diagram of a circuit configuration used as a line receiver is shown in Fig. 3.15. The input signal is attenuated by the resistive drivers R1-R2 and R8-R3. This attenuated signal is fed into a balanced dc amplifier, operating in the common base configuration. This input amplifier, consisting of Q1 and Q2, removes the *common-mode* component of the input signal. Furthermore, it delivers an output signal at the collector of Q2, which is nearly equal in amplitude to the original differential input signal. This output signal is buffered by Q6 and drives an output amplifier, Q8. The output stage drives the logic load directly.

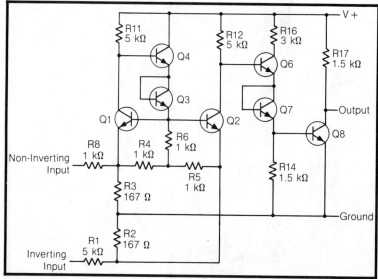

Figure 3.15—Simplified Schematic of a Typical Line Receiver

It must be noted that the line driver and line receiver circuits which have been presented are very fundamental, since the main purpose was to describe their basic functions. Today, many improved circuits are available from various manufacturers, including *line transceivers* or *line repeaters*, which will be discussed in subsection 3.4.3

3.4.2 EIA and Other Standards for Line Drivers/Receivers

A standard, by definition, is fixed. It cannot evolve with environment where it is applied. Some of them resist radical change. One example is the layout of the keys on a typewriter, which was decided by purely practical mechanical reasons but has not changed appreciably for computer keyboards. Others, however, quickly become obsolete.

In considering data transmission techniques or telecommunications in general, there are two main authorities on matters of standardization: the American Engineering Industries Association (EIA) and the United Nations' Consultative Committee for International Telegraph and Telephone (CCITT). It should be noted that while the American body actually establishes norms, the CCITT only makes recommendations, because of a conflict of interest among the member nations. As far as the norms mentioned above are concerned, both bodies are in agreement.[20]

EIA has developed several specifications to standardize the interface in data communication systems. Table 3.2 shows a comparison of various standards.[21-25] The first of these, RS-232-C, was introduced in the early 60's and has been very widely used throughout the industry. This standard was developed for single-ended data transmission over short distances and slow data rates. Today's higher performance data communication systems are rapidly making RS-232-C inadequate, with the need to transmit data faster, and over longer distances. RS-232-C's (Fig. 3.16a) interconnect length is just 15 m and its data-transmission rate 20 kbps. RS-232-C's single-ended counterpart, RS-423-A (Fig. 3.16b), allows 1,220 m between equipment and runs at data rates to 100 kbps. The CCITT's equivalents to RS-232-C and RS-423-A are the V.24 and V.10/X.26, respectively.

For data rates faster than 100 kbps over long distances, differential data transmission should be used to nullify effects of ground shifts and noise signals which appear as common-mode

Drivers/Receivers

Table 3.2 Parameters for Principal EIA Standards*

	RS-232C	RS-423A	RS-422A	RS-485
Mode of operation	single-ended	single-ended	differential	differential
Number of drivers & receivers allowed on line	1 driver 1 receiver	1 driver 10 receivers	1 driver 10 receivers	32 drivers 32 receivers
Maximum cable length (FT)	50	4000	4000	4000
Maximum data rate (bits per second)	20k	100k	10M	10M
Maximum common-mode voltage	±25V	±6V	+6V −0.25V	+12V −7V
Driver output	±5V min ±15V max	±3.6V min ±6.0V max	±2V min	±1.5V min
Driver load (Ω)	3k to 7k	450 min	100 min	60 min
Driver slew rate	30V/μsec max	externally controlled	NA	NA
Driver output short-circuit current limit	500 mA to V_{cc} or ground	150 mA to ground	150 mA to ground	150 mA to ground 250 mA to −8V or +12V
Driver output resistance (High Z state) Power On	NA	NA	NA	120k
Driver output resistance (High Z state) Power Off	300	60k	60k	120k
Receiver input resistance (Ω)	3k to 7k	4k	4k	12k
Receiver sensitivity	±3V	±200 mV	±200 mV	±200 mV

*(Reproduced from EDN by kind permission of Cahners Publishing Co., Division of Read Holdings, Inc., and the author D. Pippenger.)

voltages on the driver outputs and receiver inputs. RS-422-A (Fig. 3.16c) was defined by EIA for this purpose and allows data rates up to 10 Mbps (up to 12 m) and up to 100 kbps (for line lengths up to 1,220 m).

The shortcoming of RS-422-A is that in bus applications, multiple drivers may be required on the same data bus. The main limitations of RS-422-A in bus applications are related to the driver. When multiple drivers are connected to a common bus, only one is transmitting data and the remainder should be in a high impedance state so as not to load the bus. RS-422-A does not require the driver to be in a high impedance state except when power is off, and then only over a common-mode range from -0.25 V to $+6.0$ V. Since large positive and negative common-mode signals can appear at driver outputs connected on a bus system, it is necessary that they maintain a high impedance when disabled with power *on* and power *off* and over a wide common-mode range. Another limitation of RS-422-A in bus systems is *contention*, which is defined as more than one driver on the bus being enabled simultaneously. When this occurs, with a common-mode voltage between the drivers, appreciable current can result in excessive power dissipation and possible destruction of the driver. There are no constraints on the RS-422-A driver to protect it from destruction under these conditions. The CCITT's equivalent to RS-422-A is the V.11/X.27.

RS-485 is an enhanced version of the differential standard RS-422-A. Unlike RS-422-A, RS-485 (Fig. 3.16d) does not limit system design to a single driver/multiple receiver architecture. Indeed, its major advantage is that it permits multiple drivers, receivers and *transceivers* to operate over a two-wire bus, thus setting up a *party line* architecture. RS-485 incorporates the specifications of RS-422-A as well as specifications on the use of multiple drivers and receivers. RS-485 provides the following specifications:[21]

1. A driver's off-state output leakage current must be less than 100 mA over -7 to $+7$ V.
2. A driver must be able to produce a differential output voltage of 1.5 to 5 V under common-mode voltage conditions from -7 and $+12$ V.
3. Drivers must have internal protection from contention situations. Because multiple drivers can be connected on an RS-485 bus, two or more drivers can conceivably vie for access to one bus at the same time. The specification requires that no damage occurs to a driver under such conditions. Specifically, the standard states that a driver must be safe from

Drivers/Receivers

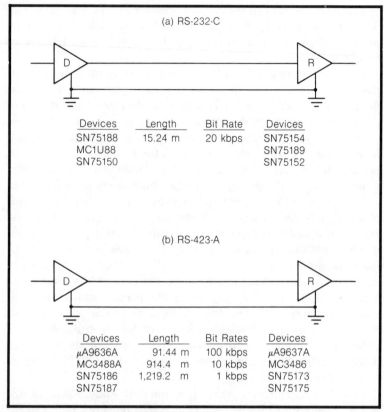

Figure 3.16a and b—Circuit Connections and Devices for EIA Standards

damage if its outputs are connected to a voltage source between -7 and $+12$ V when its input state is a logic "1" or logic "0" or if it is passive.

4. One driver can drive a total line termination resistance (R_T) of 60 Ω and as many as 32 unit loads. A unit load is based on the number of devices that can be connected on an RS-485 bus. One unit load is defined as a worst-case load that draws 1 mA under a maximum common-mode voltage-stress of 12 V. It can be based on the load of either a driver or receiver. It does not, however, include the value of line termination resistance, which can present an additional load with resistance as low as 60 Ω.

The rise-and-fall-time specs of RS-485 and RS-422-A are also different, and they impose design limitations. RS-485 drivers exhibit slower rise and fall times and therefore emit less radiation than their RS-422-A counterparts.

Transceivers/Repeaters

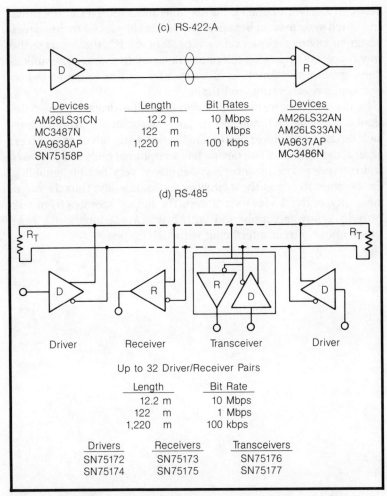

Figure 3.16c and d—Circuit Connections and Devices for EIA Standards

3.4.3 Transceivers and Repeaters

As Fig. 3.16d shows, a plurality of line driver/receiver pairs can be connected on an RS-485 party line bus. Data communications systems of this type which may have transmission lines as long as hundreds, or even thousands, of feet often experience attenuation

Drivers/Receivers

between a driver at one end of the line and a receiver at the other end. Such systems can benefit from the use of special *transceivers* designed as *bus repeaters*. Bus repeaters are ICs that operate the way repeaters on analog telephone lines do. In a data communications system, the function of a repeater can be summarized as reshaping, regenerating and timing.[24]

Figure 3.17 shows a dual differential line driver/receiver designed for use in industrial, noisy environments. Applications include use as a differential, or single-ended line driver or receiver, or as a line repeater. The device has an optional built-in *hysteresis* and reference. The hysteresis capability is very flexible, enabling the designer to adjust the switching thresholds, and thus the noise immunity, of the device as required. The device operates from a 12 V to 15 V power supply and with low current inputs (0.4 mA) which allows direct interfacing with CMOS.

Transceivers/Repeaters

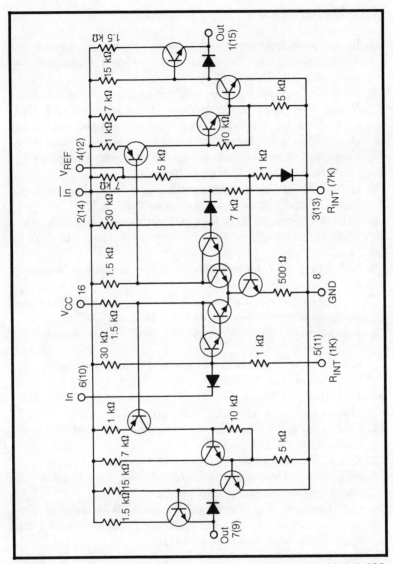

Figure 3.17—Dual Differential Driver/Receiver (Repeater) Model 396 (Courtesy of Teledyne Semiconductor)

3.5 References

1. R. K. Southard, *Interconnections System Approaches for Minimizing Data Transmission Problems, Computer Design*, March 1981, pp. 107–116.
2. T. Balph, *Implementing High Speed Logic on Printed Circuit Board*, Wescon/81, San Francisco, CA, September 15–17, 1981.
3. J. R. Benham, *Crosstalk and Common Mode Aspects of Differential Line Driver and Receiver Circuits*, EMC '82— International Symposium on Electromagnetic Compatibility, Santa Clara, CA, September 8–10, 1982.
4. M. Davidson, *Understanding the High Speed Digital Logic Signal, Computer Design*, November 1982, p. 79.
5. E. E. Davidson and R. D. Lane, *Diodes Damp Line Reflections Without Overloading Logic, Electronics*, February 19, 1976, p. 123.
6. N. L. Bragg, *ECL for the TTL Designer, Electronic Engineering*, October 1980, p. 41.
7. C. J. Georgopoulos, *Squelching Noise in Instruments and Systems, Machine Design*, July 22, 1971, pp. 74–79.
8. J. B. Marshal, *Flat Cable Aids Transfer of Data, Electronics*, July 5, 1973, p. 89.
9. E. Trompeter, *Noise in Cable Systems*, Trompeter Electronics, Inc., Chatsworth, CA, July 1, 1969.
10. J. H. Hull, *Characterization of Transmission Media, Telecommunications*, February 1984, pp. 70–72.
11. C. J. Georgopoulos, *Fiber Optics and Optical Isolators*, Gainesville, VA, DWCI, 1982.
12. J. D. Kraus, *Antennas*, McGraw-Hill, New York, 1950.
13. W. S. Bennett, *Controlling Radiated EMI from Digital Systems, RF Design*, March 1985, p. 24.
14. U. S. Federal Communications Commission *Methods of Measurement of Radio Noise Emissions from Computing Devices, Rules and Regulations*, 1981.
15. R. K. Keenan, *Digital Design for Interference Specifications*, Vienna, VA, The Keenan Corp., 1983.
16. M. V. Brunt, *The Design of EMI Hardened Connectors, RF Design*, May 1985, pp., 29–31.
17. H. Hazzard and R. Kiefer, *EMI Reduction Starts With Connector Doubling as Low-Pass Filters, Electronic Design*, March 14, 1985, p. 181.

18. E. VanderHeyden and J. H. Bogar, *Designing for Compliance With FCC EMI Regulations*, January 1982, p. 189.
19. R. J. Widlar and J. J. Kubinec, *Integrated Circuits for Digital Data Transmission, Application Note AN-22*, National Semiconductor Corp., February 1969.
20. Elektor Editorial Staff, *The RS 423 Interface, Elecktor Electronics 109*, May 1984, pp. 5.65—5.66.
21. D. Pippenger, *ICs Extended RS-422 to Multistation Applications, EDN*, March 21, 1985, p. 181.
22. Texas Instruments, *Line Circuits for IEEE-488 and EIA Industry Standards, Application Note CL-558*.
23. R. Burgess, *RS-422 and Beyond, Electronic Engineering*, October 1981, p. 81.
24. M. K. Kimyacioglu, *Implement Bell T1C PCM Repeater Using Just Two ICs, EDN*, March 1985, pp. 195–200.

General References

- Engineering Staff/Don White Consultants, Inc. *The Role of Cables and Connectors in Control of EMI*, EMC Technology, Vol 1, No. 3, July 1982, p. 16.
- C. R. Paul and D. Koopman, *Sensitivity of Crosstalk in Balanced Twisted Pair Circuits to Line Twist*, EMC '83—International Symposium on Electromagnetic Compatibility, Zurich, Switzerland, March 1983.
- J. Williamson, *Metal Cable: Backbone of Telecomms Network, Communications Engineering International*, April 1984, p. 25.
- R. A. Crouch, *Miniature Coaxial Cable: State of the Art, Electronic Packaging and Production*, February 1982, p. 99.
- A. G. Schuh, *Fundamentals of Wiring and Cabling, Electronic Packaging and Production*, January 1983, p. 242.
- T. Hopkins, *Line Driver and Receiver Considerations, Application Note AN-708A*, Motorola Semiconductor Products, Inc., Phoenix, AZ, 1978.
- Texas Instruments, *Linear and Interface Circuits Seminar Bulletin*, January 1979.
- G. F. Dooley, *Line Circuit Interfaces for a Digital Switching System, IEEE Transactions on Communications*, Vol. COM-27, No. 7, July 1979, pp. 978–982.

Drivers/Receivers

- L. Silverman, *ICs Control Major Elements—and Costs—of 3-State Modem Transmitters, Receivers, Electronic Design* 19, September 13, 1979, pp. 134–137.
- R. Sproul, *Trouble-Free Interfacing of TTL and RF Components, Microwaves & RF*, October 1983, p. 121.
- S. Goodspeed and M. L. Baynes, *Speedy CMOS Logic Drives Transmission Lines With Schottky Efficiency, Electronic Design*, May 16, 1985, pp. 211–216.

Chapter 4
Terminal and Peripheral Interfaces

Along with an increased volume of digital data processed by business has come an increased demand for interfaces suitable for terminal-to-terminal, terminal-to-computer and computer-to-peripheral transmission of data. Such transmission, however, is not free from interference problems, and special techniques are required to reduce them.

4.1 Introduction

In the increasingly sophisticated world of data transmission, the use of standard and special interfaces is becoming more important. By defining the active and passive elements of the communication links between CPUs, computers and associated terminals and peripherals, industry standards have brought form, function and order to a previously chaotic system, eliminating delays, reducing costs and improving productivity.

Each terminal or peripheral interface to be used should be compatible with the available equipment and should comply to certain standards and regulations, depending on the application and environment in which it will operate. The environment is responsible for many interference problems which must be seriously considered by the designers of interface circuitry and systems integrators. Some interference problems addressed in this chapter include: noise suppression in computer grade connectors and peripheral actuators, conductive shield termination, use of optocouplers and EMP radiation detector circuitry.

Terminal/Peripheral Interfaces

4.2 Terminal Interfaces: Some Definitions and Standards

While terminals have been physically adapted to the user in many ways, functional flexibility to suit user preference and convenience has become available recently through the availability of new interfaces and standards. Today's low-cost intelligent terminals incorporate the adaptability users need to define functions that were previously available only at the host computer.

4.2.1 Some Basic Definitions

1. In data communication, a *terminal* usually is said to be any unit of hardware that interfaces with, and provides service to, a person using it. Its primary goal is to provide access for the user to some desired function or information. Terminals range from simple devices with little "intelligence" to large computers that, at times, can provide a simple terminal function which is far below the capability of its primary function.[1] Transfer of data between a computer and a public network, or in-plant installation, is carried out by a *data terminal equipment (DTE)*, which is part of the computer system, and a *data circuit terminating equipment*, (DCE), popularly called a *modem*,[2] which is connected to a transmission line (Fig. 4.1).
2. *Modem* is a contraction of the terms *mod*ulator and *dem*odulator. Its alternate name, in data communication systems, is data circuit terminating equipment (DCE) as was mentioned above.

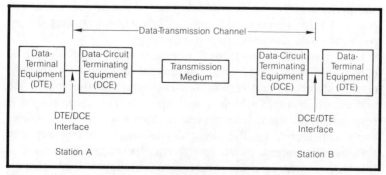

Figure 4.1—Typical Data-Transmission System

3. *CRT Terminal* is a terminal device, and its main function is to convert stored, compactly coded characters located in memory into a repetitive form that a *cathode ray tube (CRT)* can understand and use.

4.2.2 Common Data Interfaces for Digital Communications Links

The rapid development of wideband transmission technologies (including satellite, lightwave and digital radio), the impact of digital switching and the improved performance on wire and coaxial systems have contributed to the rapid evolution of digital networks. The deregulatory environment in general, and the Bell System divestiture in particular, has encouraged the development of these new technologies and services. Testing methods used on digital communications links must accommodate data rates from a few hundred bits per second (bps) to several hundred million bps. Therefore, compatibility with a wide range of physical interfaces is required. Table 4.1 summarizes important characteristics for many of the common data interfaces in use.[3]

Terminal/Peripheral Interfaces

Table 4.1 Common Data Interface Characteristics*

Interface	Connector Type	Data Rate or Speed Range[5]	Interface Timing Mode	Line Code	Signal Format	Nominal Signal Levels[6]	Nominal Termination or Load Impedance
EIA RS-232[1]	25 pin "D" type	to 20 kbps	Synchronous or Asynchronous or Isochronous	Not Specified	Unbalanced bipolar	±5 to ±25 V	3 to 7 kΩ
EIA RS-449[2]	37 pin "D" type with 9 pin "D" type for sec. channels	Unbalanced to 100 kbps Balanced to 10 Mbps	Synchronous or Asynchronous or Isochronous	Not specified	Unbalanced bipolar and/or Balanced	Unbalanced: ±4 to 6 V Balanced: ±2 to 6 V	Unbalanced >400 Ω Balanced >400 Ω w/optional 100 Ω termination
CCITT V.35[3]	34 pin rectangular	to 10 Mbps	Synchronous or Asynchronous or Isochronous	Not specified	Data & clock: Balanced bipolar Signaling: Unbalanced Bipolar	Data & clock: ±0.55 V Signaling: ±5 to 25 V	Data & clock: 100 Ω Signaling: 3 to 7 kΩ
WECO 303[4]	Multiple coaxial	to 500 kbps	Synchronous or Asynchronous or Isochronous	Not specified	Current mode	MK < 5 mA SP > 23 mA	100 Ω
CCITT G.703	Not specified	2.048 Mbps	Isochronous	HDB3	Balanced bipolar pseudo-ternary or Unbalanced bipolar pseudo-ternary	Balanced: 3 V or Unbalanced: 2.37 V	Balanced: 120 Ω or Unbalanced: 75 Ω

Digital Communications

	Connector	Rate	Format	Code	Interface	Voltage	Impedance
DS1	15 pin "D" type or WECO 310 or Bantam	1.544 Mbps	Isochronous	AMI or B8ZS	Balanced bipolar pseudo-ternary	±3 V[8]	100 Ω
DS1C	WECO 310 or Bantam	3.152 Mbps	Isochronous	AMI	Balanced bipolar pseudo-ternary	±3 V[8]	100 Ω
DS2	WECO 310 or Bantam	6.312 Mbps	Isochronous	B6ZS	Balanced bipolar pseudo-ternary	0.2 to 7.3 dBm[9]	110 Ω
DS3	BNC or WECO 440A	44.736 Mbps	Isochronous	B3ZS	Balanced bipolar pseudo-ternary	−1.8 to 5.7 dBm[10]	75 Ω

1. The combination of CCITT V.28 and V.24, along with ISO 2110, produces a similar and compatible interface. EIA RS-449 refers to EIA RS-442A and RS-423A for electrical characteristics. The combination of FED STD 1031, 1020A and 1030A or of ISO 4902, and CCITT V.24, V.10, and V.11 produce a similar and compatible interface.
2. CCITT V.28 and ISO 2593 are used along with CCITT V.35 to completely define the interface.
3. WECO 303 specifies that electrical characteristics per EIA RS-232 be used for the Data Terminal Ready and Ring Indicator signaling leads.
4. Upper limits are nominal; operation at higher rates is sometimes possible.
5. Signal levels are specified at the driver outputs. For balanced operation, peak lead-to-lead voltages are specified.
6. Although connectors are not specified, BNC and WECO 310 connectors are commonly used for unbalanced and balanced operation respectively.
7. Levels are specified at the DSX point.
8. This power level is specified at the DSX2 point when an all 1's pattern is transmitted and the measurement is made in a 2-kHz band about 3.156 MHz.
9. This power level is specified at the DSX3 point when an all 1's pattern is transmitted and the measurement is made in a 2-kHz band about 22.368 MHz.

*(Reprinted by permission from Telecommunications, © 1984 by Horizon House—Microwave, Inc.)

Terminal/Peripheral Interfaces

4.2.3 International Modem Standards

Many modems, especially older ones, use fairly straightforward frequency, or phase shift, modulation techniques to encode digital data onto the analog voice bandwidth of the telephone system. Simple asynchronous (start/stop) is so easy to implement in silicon that several firms offer multimodem chips combining both the Bell and CCITT schemes for 300 and 1,200 bps—the Bell-103/CCITT V.21 and Bell-202/CCITT V.23, respectively. Traditional data processing machines are more likely to use synchronous communication at speeds anywhere between 2,400 and 9,600 bps.

When using *differential phase shift keying (DPSK)* and *quadrature amplitude modulation (QAM)* techniques, as used in the 2,400, 4,800 and 9,600 bps modems (outlined by CCITT V.26, V.27 and V.29), they can only support data in one direction at a time (*half-duplex*) on the limited 300 to 3,400 Hz bandwidth on the public network.

For *full-duplex* operation, the Bell-212A was developed in the United States. After some amendments, it became the CCITT V.22 recommendation, using *frequency shift keying* (FSK) modulation. V.22 supports both synchronous and asynchronous data at the digital interface at 1,200 bps, but actually transmits its analog data synchronously. The speed of V.22 can be doubled by adding amplitude modulation on top of the phase shifts. The new scheme is V.22 *bis* and is included in the latest CCITT recommendations. It allows for 2,400 bps full-duplex transmission over two-wire dialed or leased lines and handles both synchronous and asynchronous digital data.

A competitor to V.22 *bis* is V.22 *ter* which initiates a new role with its complex data encoding and modulation. Only one DSPK channel is imposed on the telephone line and the two-bit groups are represented at 90° phase changes. However, the modem remembers the data it has sent and subtracts the echo from the incoming signal, leaving only the data transmitted by the remote modem.

Another user of echo cancellation is V.32, the 9,600 bps, two-wire, full-duplex specification. Groups of four bits are encoded into 16 QAM states, as in V.22 bis. Table 4.2 summarizes various international modem standards and their characteristics.[4]

To meet the growing demand for integrated voice and data communications, AT&T Information Systems has proposed the

Modem Standards

Table 4.2 International Modem Standards*

CCITT recommendations	Speed in bits/s		Data	Modulation	Connection		Bell modems	CCITT/Bell compatibility
	Normal	Fallback			Two-wire	Four-wire		
V.21	0 to 300	—	A	FSK	full	—	103/113	No
V.22	1200	600	A/S	DPSK	full	—	212A	Yes, 1200 bits/s mostly
V.22*bis**	0 to 300	—	A	DPSK	full	—	—	No, 300 bits/s
V.23	2400	1200	A/S	QAM	full	full	—	—
V.26	1200	600	A/S	FSK	half	full	202	Similar at 1200 bits/s
V.26*bis*	2400	—	S	DPSK	—	full	201	Yes, 2400 bits/s V.26 Alt B
V.26*ter*	2400	1200	A/S	DPSK	half	full	—	—
V.27	4800	—	S	DPSK	full	—	—	—
V.27*bis*	4800	2400	S	DPSK	—	full	208	No
V.27*ter*	4800	2400	S	DPSK	half	—	208	No
V.29	9600	7200/4800	S	QAM	half	full	208	No
V.32*	9600	4800/2400	A/S	QAM	full	—	209	No

Data: A = asynchronous; S = synchronous.
Modulation: FSK = frequency shift keying; DPSK = differential phase shift keying; QAM = quadrature amplitude modulation.
*Recommendations formulated since 1980.
(From "Modern Standards Aimed at Modem Transmission Schemes," by M. Repko, p. 120. Copyright by Computer Design, August 1985. All rights reserved. Reprinted by permission.)

Terminal/Peripheral Interfaces

digital multiplexed interface (DMI) standard for data communication (Fig. 4.2). The DMI would provide up to 24 channels, each operating at 64 kbps using the North American 1.544 Mbps digital line rate possible with AT&T's existing network. Figure 4.2 indicates three alternatives for data transmission between a PBX and a host computer, including modems, on a per-channel basis operating at a maximum data rate of 19.2 kbps, product-specific data modules which provide two 64 kbps data channels (30 in Europe), each with 64 kbps data rates. Figure 4.3 displays the relative cost per number of channels for the above three categories of data transmission. The DMI is compatible with *integrated services digital network (ISDN)* and with digital *private branch exchanges (PBX)* currently being installed at a rate of about 10,000 per year in the United States.[5]

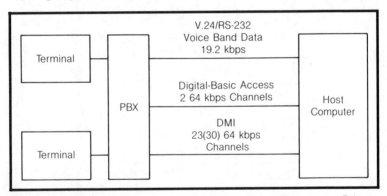

Figure 4.2—Three Alternatives for Data Transmission between a Private Branch Exchange (PBX) and a Host Computer

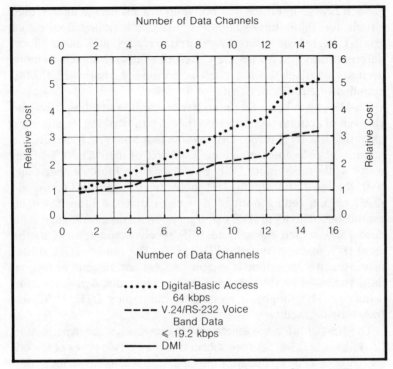

Figure 4.3—Relative Cost Per Number of Channels Shown in Fig. 4.2

4.2.4 The EIA RS-449-A Data Communications Interface[6]

The EIA RS-449-A interface can be incorporated in a data communications equipment to achieve data-transmission rates to 2 Mbps and interface-cable lengths to 1,200 m. These specs, combined with RS-449's differential driver/receiver configuration and expanded control-signal set, allow RS-449 to overcome the drawbacks of RS-232-C.

Like all complete physical-interface standards, RS-449 provides the ability to establish, maintain and disconnect a physical path to a transmission medium connecting data-terminal equipment (DTE), such as a modem or digital transmission service unit. To understand how to implement the interface, one should consider functional, procedural, electrical and mechanical aspects that are summarized below:

Each RS-449 interface function employs a separate interchange circuit for data, timing, control, diagnostic loops and signal ground. A set of interchange-circuit names and mnemonics, different from that for RS-232-C, describes RS-449 signals to more accurately reflect the functions performed. Additionally, RS-449 includes 10 circuits not found in RS-232-C.

RS-449 procedures for interchange circuits are based on the concept of action/reaction pairs (that is, handshaking signals). For example, the DTE might turn *remote loopback (RL)* circuit (the action) *on* to request a remote loopback of data. The DTE then waits while the remote DCE establishes the loop. At that time, both the local and remote DTE activate (turn on) the *test mode (TM)* circuit, telling both DTE devices that the loop has been established. The local DTE can transmit data on interchange *send data (SD)*, which the remote DCE receives and returns to the local DTE on *receive data (RD)* circuit. The remote DTE, recognizing that a remote loop is established, cannot transmit or receive data. Deactivating the loopback involves a similar action/reaction sequence. The loopback requires a full-duplex DTE, DCE and transmission facility.

The RS-449 interface electrical characteristics are described in RS-423-A and RS-422-A (see subsection 3.4.2). Two types of RS-449 electrical-interface configurations exist. One uses RS-423-A generators on all interchange circuits; it is interoperable with RS-232-C (at data rates to 20 kbps and distances to 15 m). The other provides a single high-performance interface capable of speeds up to 2 Mbps.

Two connectors can be used to implement RS-449. A 37-pin device furnishes the basic interface to facilitate one primary serial communication channel, and a 9-pin unit serves as an optional secondary channel. The international equivalent of these 37/9-pin connector standards is ISO 4902. The small size of the DTE connector permits compact mounting arrangements for multiple connectors while assuring adequate clearness. The pin arrangement plan minimizes crosstalk and facilitates the design of an adapter for interoperation with RS-232-C.

4.3 Terminal Interfaces: Representative Design Examples

The vast majority of the terminal market is filled with terminals with two basic capabilities:[7]

1. Keyboard capability of *generating*, and a monitor capabable of *displaying*, a full alphanumerical character set
2. Electronics capable of *sending* and *receiving* data via a communication line (transmission medium) to and from a remote host computer (usually a serial line)

The interface to be used should be compatible with the available equipment and should comply to certain standards and regulations, depending on the application and the environment it will operate.

4.3.1 Man/Machine Interface[8-11]

Conventional *man/machine* interfaces for industrial control and measurement systems have included switches, knobs, dials, analog-to-digital converters, lights, meters and readouts with each component being dedicated to one parameter or function.

Computers and CRTs with typewriter keyboards have added a new dimension to the operator interface, and keyboard/display units have augmented control and measurement panels. On the other hand, *touch* technology simplifies further man/machine interface. It includes conductive membrane, capacitive overlay and scanning infrared beam through system designs.

Compared to other devices available to facilitate communication between user and computer, touch has some special advantages. One is that it provides a fast, easy and natural way to control computer operations without special input tools and with limited computer experience. Conceptually, a touch system can be divided into two major components: the sensor unit and the control unit. The principal differences among touch systems lie in the sensor technology membrane, capacitive or infrared (Table 4.3).[11] The control unit is typically microprocessor-based and includes the electronics needed to interface the sensor unit, and the firmware to process touch activations to communicate them to the host.

Display devices can be either *CRTs* or *plasma displays (flat type)*. Intensive R&D efforts are now being directed at new flat-planel technologies, such as *liquid crystal displays (LCD)* and *electroluminescent (EL)* displays which, so far, are not as attractive as CRTs.

Terminal/Peripheral Interfaces

Table 4.3 Touch Technology Comparison*

	Membrane	Capacitive Overlay	Infrared Scanning
ADVANTAGES	• Potential for better resolution • Little Parallax • Relatively inexpensive	• Little parallax • More transparent than resistive	• Unobstructed viewing path, nonglare screen • Most reliable (solid state, 1% duty cycle on leds) • Operates in a variety of environmental conditions • Scratches, dirt, oil, etc. do not cause malfunction • Failed unit customer-repairable • Different operating modes allow optimum programming capability • Resolution as fine as that of display
DISADVANTAGES	• Conductive material in optical view path degrades screen image • Reflective surface (mylar-glass) increases glare • Vulnerable to vandalism (factory floor, hotel lobby) • Failed units are not repairable	• Capacitive material in optical viewing path • Potential problem from moisture and static of user's hands • Scratches may destroy capacitive layer • Pencil or gloved hand may not work • Must be continually calibrated	• Higher cost • Requires more space in front of video display

(*Reprinted by permission of the publisher from the Winter 1984 edition of COMPUTER REVIEW®. Copyright © 1984 by West World Productions, Inc.)

4.3.2 DCE-to-DTE Interconnections with Balanced/Unbalanced Circuits

EIA interfaces can be used when implementing DCE-to-DTE interconnections. As mentioned previously (see subsection 4.2.4), the RS-449-interface electrical characteristics are described in RS-423-A and RS-422-A. RS-423-A (CCITT V.10/X.26) specifies unbalanced operation on interchange circuits and is interoperable with both RS-232-C and RS-422-A, facilitating the evolution of an existing RS-232-C interface. With RS-423-A, signaling occurs over one wire per interchange circuit with a single common-return for a reference voltage. RS-422-A (CCITT V.11/X.27) specifies balanced operation, employing differential signaling over a pair of wires for each interchange circuit.

Both RS-232-C and RS-422-A use a balanced differential receiver. RS-423-A's common return connects to signal ground only at the generator end; RS-422-A's two generator signal wires connect to the receiver either directly or through an optional terminating resistor. Figure 4.4 illustrates the grounding arrangements for balanced (RS-422-A) and unbalanced (RS-423-A) generators with differential drivers.[6]

Terminal/Peripheral Interfaces

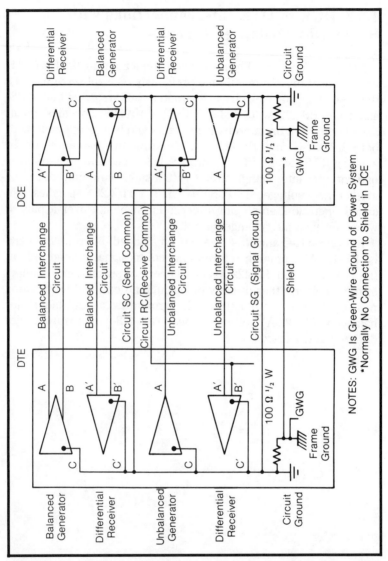

Figure 4.4—DCE/DTE Interconnection Diagram Illustrating Ground Arrangements for Balanced and Unbalanced RS-449 Interchange Circuits (Reproduced from *EDN* by Permission of Cahners Publishing Company, Division of Read Holdings, Inc. and Author A.J. Aveissberger)

4.3.3 CPU-to-Remote CRT Terminal Interface

Figure 4.5 shows an interconnection scheme between a central processor unit (CPU) and a remote CRT terminal, with other intermediate device connections, including a repeater.[12] This is a typical *bus* application using a combination of drivers, receivers and transceivers according to RS-485 (see also subsection 3.4.2). Since the various types of bus systems will be discussed in Chapter 5, only a brief description of the above system will be presented here.

The *bus line*, where up to 32 driver-receiver pairs can be connected, is terminated at each end by a resistor matched to the characteristic impedance. Typically, this will be 100 or 600 Ω. The stubs on the line connecting the various drivers and receivers to the main bus can be placed anywhere along its length, but they must be kept short (less than 0.3 m) to prevent reflections on frequencies of 1 MHz and above. Length restrictions are circumvented by bidirectional repeaters.

Terminal/Peripheral Interfaces

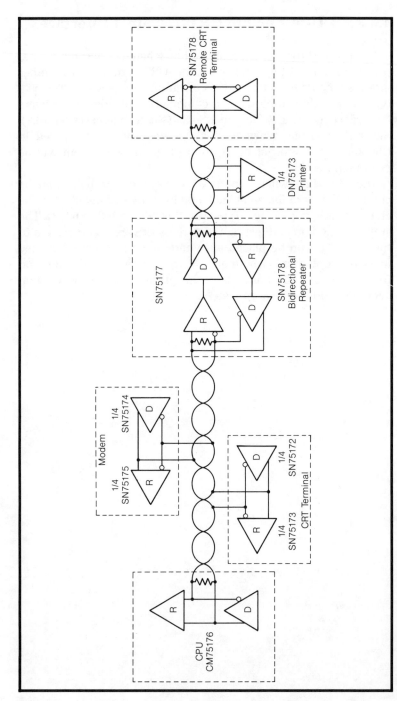

Figure 4.5—CPU-to-Remote CRT Terminal Interface and Other Devices with Bus-Line Bidirectional Repeater (Reproduced by Permission of *Electronic Engineering*, © Morgan-Grampian Publishers, Ltd., and Author R. Burgers)

4.4 Peripheral Interfaces

Finding interfaces that are relatively low in construction cost and capable of a high degree of configuration flexibility is the general intent of systems integrators. Standard peripheral products offered by a host computer's manufacturer need not limit a system's capabilities. By incorporating a capable peripheral controller, a computer system can utilize peripherals other than those normally supported by the host computer's manufacturer. Such controllers emulate a peripheral interface which is supported by the computer manufacturer.

4.4.1 Some Useful Definitions

- *Peripheral Equipment* is equipment that is external to and not a part of the central processing instrumentation. This would include such equipment as tape punches and readers, magnetic tape or disk storage units, graphic recorders, etc.
- *Magnetic Disk Storage* is a storage device or system that consists of magnetically coated disks, on the surface of which is stored information in the form of magnetic spots arranged in a manner to represent binary data.
- *Magnetic Tape Storage* is a storage system using magnetic spots (bits) on metal or coated-plastic tape. The spots are arranged to read out in the desired code as the tape travels past the read-write head.

4.4.2 Peripheral Interface Chips[13]

In order to take advantage of microprocessor advancements, logic and memory had to improve. For example, an entire second generation of high speed logic appeared in the 74CXX/74SCXX products. These logic devices offered low-power Schottky transistor-transistor logic (LS/TTL) equivalent propagation delays at complementary metal oxide semiconductor (CMOS) operating power levels, and provided the necessary high-speed "glue" for advanced CMOS microprocessor systems (see also Chapter 2). Similarly, higher density CMOS memories are now available at much greater speeds. Functional options include both synchronous and asynchronous memory operation to meet specific system requirements.

Terminal/Peripheral Interfaces

Various sophisticated CMOS microprocessors have brought high performance to low-power designs; however, they cannot reach their full potential without equally high performance, lower power consumption support chips. For the most part, CMOS peripheral circuit design efforts have lagged. This has somewhat limited the development of systems attempting to use CMOS devices exclusively. A new family of microprocessor peripheral circuits (Table 4.4) fills this void by providing increased performance and functionality without sacrificing low power consumption.

The 80C86 peripheral product line (Table 4.4) from Harris Corp. has a wide functional range, so that complex, high-performance systems for low-power applications can be designed. The peripherals are TTL compatible CMOS versions of industry standard NMOS devices. In addition, they incorporate improvements that eliminate traditional problems in hardware, software and power consumption.

Table 4.5 shows that the peripheral family's architecture is fully compatible with 80C85- and 80C86-type microprocessors. However, the popular two-line control method for data movement, using *read (RD)* and *write (WR)* lines, allows interface to almost any recent generation microprocessor. Dual specifications for the logical 1 output voltage (V_{OH}) ensure interface compatibility of peripherals with both CMOS and TTL devices. Even in an all-CMOS design, there may be the need for circuit functions available only in NMOS, or bipolar. In this case, the peripherals allow direct interface without pullup resistors or additional circuits of other technologies. Thus, the retrofit problems that occur with non-TTL compatible devices are eliminated.

Table 4.4 CMOS 80C86 Family Peripheral Support Chips*

Part	CMOS device type/function	Comments
82C82	octal latch	T_{PD} = 35 ns max at C_L = 300 pf
82C84A	clock generator/driver	8-MHz system clock frequency
82C88	bus controller	Status decode function
82C54	programmable interval timer	10-MHz count frequency
82C55A	programmable peripheral interface	Control word read capability
82C59A	priority interrupt controller	8 user defined priority interrupt requests

*From "Providing CMOS Benefits to Peripheral Chips," by W. J. Niewierski, p. 124. Copyright by Computer Design, 1983. All rights reserved. Reprinted by permission.

Table 4.5 Peripheral Interface Compatibility*

	Logical 1 Input voltage (VIH)	Logical 0 Input voltage (VIL)	Logical 1 Output voltage (VOH)	Logical 1 Output current (IOH)	Logical 0 Output voltage (VOL)	Logical 0 Output current (IOL)
80C86	2.0 V/2.2 V Ind/Mil	0.8 V	3.0 V	−2.5 mA	0.4 V	2.5 mA
Peripherals			V_{cc} − 0.4 V	−100 μA		
NMOS 8086	2.0 V	0.8 V	2.4 V	−400 μA	0.45 V	2.5 mA
Family CMOS	70% V_{cc}	30% V_{cc}	V_{cc} − 0.5 V	−10 μA	0.4 V	2.0 mA
LS/TTL	2.0 V	0.8 V	2.5 V	−400 μA	0.4 V	4.0 mA—Military
			2.7 V		0.5 V	8.0 mA—Commercial

*From "Providing CMOS Benefits to Peripheral Chips," by W. J. Niewierski, p. 124. Copyright by Computer Design, 1983. All rights reserved. Reprinted by permission.

4.4.3 Disk and Tape Drive Interfaces

New *disk* and *tape* drive interfaces usually result from extensions made to existing interfaces, with modifications geared to new requirements. The major disk drive interfaces, including SA450, ST506/412, ESDI, SMD, SASI, SCSI and IPI, are shown in Fig. 4.6.[14]

In the category of small disk drives, the most recent device interface is the *enhanced small device interface (ESDI)* for high-performance 5 1/4-inch *rigid disk* drives. The SA450 is the most popular 5 1/4-inch *floppy disk* interface, and has been adopted by the new generation of 3 1/2-inch floppy disks. Floppy disk drives offer various electrical connections, called jumper options, which allow a system integrator to configure a disk drive to meet a wide range of user requirements and cost considerations.[15]

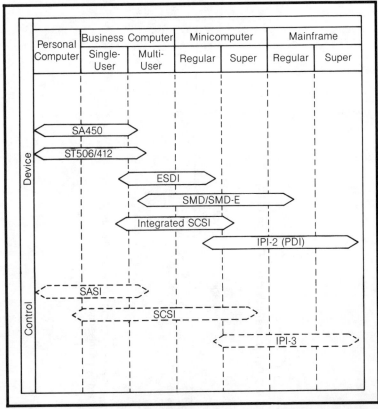

Figure 4.6—Disk Drive Interfaces Span Wide Performance Range (Reproduced from *Mini-Micro Systems* by Permission of Cahners Publishing Co., Division of Read Holdings, Inc. and Author I.D. Allan)

The ST506 interface is the *de facto* standard for low-performance *Winchester disk drives*. An improved version of this interface is the ST412, providing buffering of the step counts. The ESDI extends the ST506/412 interface and has step and serial modes of operation on disk, plus a tape mode.

The *storage module drive (SMD)* interface incorporates the data separator and is the only interface that is appropriate for disk drives from 5 1/4 inches through 14 inches. SMD is the *de facto* standard on high-performance Winchesters, and only recently became available on high-capacity 5 1/4-inch disks. SMD-E is an enhanced version of SMD which provides more diagnostic and error-reporting capabilities. The *intelligent peripheral interface (IPI)* will gradually replace the SMD interface, mainly because the IPI can accommodate higher transfer rates.

In the late 1970's Shugart developed the *Shugart Associates Systems Interface (SASI)* which became the preferred interface for OEMs. Various changes resulted in the *small computer systems interface (SCSI)* which provides more functionality.

Returning to the IPI interface, it is defined in four levels. Levels 0 and 1 make up the physical interface. The physical interface, plus one of the two command sets, defines either an IPI-2 (level 2) or IPI-3 (level 3) interface. IPI-2 allows smaller disk gaps than are presently in use, thus increasing usable data storage. IPI-3 is designed for the software programmers who need to know only the generic characteristics of peripherals and who want to provide commands in a language not unlike that used by a high-level compiler.

On the other hand, *streaming tape* drives—both 1/2 and 1/4 in.—were conceived as backup devices for Winchester disks. In fact, the term streaming has become virtually synonymous with this backup function. Responding to the demands for standardization, the working group for *quarter-inch tape cartridge compatibility (QIC)* was formed. QIC spent its original efforts developing three standards: QIC-02, QIC-24 and QIC-36.[16] The three standards cover drive, media, and control compatibility and have helped establish *cartridge tape* as a viable product. QIC-02 specifies the host intelligent interface—the CPU's pipe to drive. QIC-24 defines the format in which data are recorded on the tape. QIC-36 defines the peripheral end of the interface. Functions at this end include tape control and phase locked loop design of the drive. The QIC group is continuing its study on new standards.

Terminal/Peripheral Interfaces

4.4.4 Interface for Micro-to-Mini Data Transfers

Many users of small computer systems have recognized the need for streaming tape drives as backup but find it difficult to connect them to their standard interface. One solution is to use a specifically designed *streamer buffered interface (SBI)* board which would provide four standard interfaces:[17] an RS-232-C (V.24) compatible serial link, a 20 mA serial link, an 8-bit Centronics parallel and a full IEEE 488 compatible.

In attempting to meet the requirements of the many different microsystems currently available, the buffered interface has to be configurable or readily redesigned to provide a number of different standard interfaces. Newer designs are attempting to cope with this need by adapting the modular approach shown in Fig. 4.7. The SBI board illustrates how the design can lend itself to rapid adaptation to meet the requirements of a new interface.

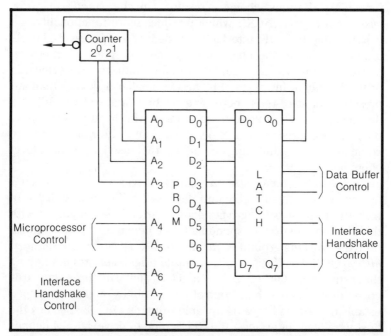

Figure 4.7—Streamer Buffered Interface (SBI) Board (Reprinted by Permission of the Publisher from the Winter, 1984, *Computer Review*, © 1984 by West World Productions, Inc.)

4.4.5 Peripheral Interface Controllers[18-20]

All peripheral devices require controllers. The controller's function is to manage the peripheral device, and to interface the peripheral to the computer's system bus for *input/output (I/O)*. Specialized controllers exist, such as *direct memory access (DMA)* controllers for I/O and channel controllers, as an intermediate between the computer and groups of peripherals.

Currently, peripheral-controller products available to system integrators can provide equivalent or higher performance than a computer manufacturer's device with the advantages of lower cost, smaller size, higher built-in intelligence, component/interconnect count reduction and peripheral-device selection flexibility.

More important is the trend to intelligent peripheral controllers. The intent of an intelligent controller is to distribute the processing load and reduce the demands on the CPU. Use of an intelligent controller also greatly simplifies the complexity of operating system software and reduces the interrupt handling load of the CPU. Programmability is very important to interface peripherals with a variety of host CPU's and to build in the intelligent processing that afloats the host.

Figure 4.8 shows a block diagram of a hard disk controller. Other peripherals are handled in a similar manner. The basic micro machine (sequencer, microcode, microprocessor) provides the intelligence of the controller data from the disk, is buffered in a small *FIFO (first-in, first-out)*, examined by the AM9520 *BEP (burst error processor)* and then put on the controller bus. Since this is an intelligent device, the data is deposited in the RAM buffer until a complete record or multiple records are transferred. The AM2900 processor can control any error correction in conjunction with BEP. For lower data rates, the AM2900 processor could be microcoded with the error correction algorithm, thus eliminating the BEP.

When processing is complete the AM2940 DMA and AM2950 I/O port allow transmission of data to the CPU at over 5 million words per second (wps). The AM2914 should be used in a controller design to handle all of the interrupts generated, both by the peripheral and the host computers. The AM2914's prioritizing function allows the interrupts that need faster response to be handled first.

Terminal/Peripheral Interfaces

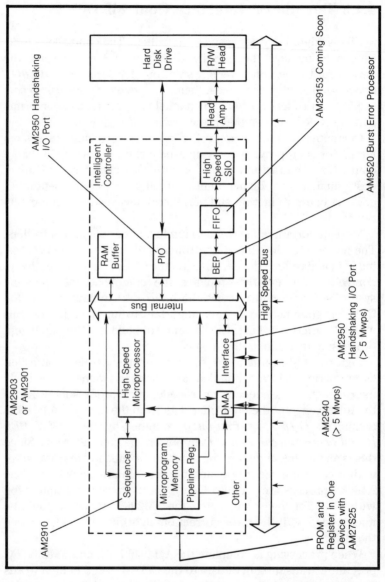

Figure 4.8—Block Diagram of Hard Disk Controller *(Courtesy of Advanced Micro Devices, Inc.)*

The ST506 interface is the *de facto* standard for low-performance *Winchester disk drives*. An improved version of this interface is the ST412, providing buffering of the step counts. The ESDI extends the ST506/412 interface and has step and serial modes of operation on disk, plus a tape mode.

The *storage module drive (SMD)* interface incorporates the data separator and is the only interface that is appropriate for disk drives from 5 1/4 inches through 14 inches. SMD is the *de facto* standard on high-performance Winchesters, and only recently became available on high-capacity 5 1/4-inch disks. SMD-E is an enhanced version of SMD which provides more diagnostic and error-reporting capabilities. The *intelligent peripheral interface (IPI)* will gradually replace the SMD interface, mainly because the IPI can accommodate higher transfer rates.

In the late 1970's Shugart developed the *Shugart Associates Systems Interface (SASI)* which became the preferred interface for OEMs. Various changes resulted in the *small computer systems interface (SCSI)* which provides more functionality.

Returning to the IPI interface, it is defined in four levels. Levels 0 and 1 make up the physical interface. The physical interface, plus one of the two command sets, defines either an IPI-2 (level 2) or IPI-3 (level 3) interface. IPI-2 allows smaller disk gaps than are presently in use, thus increasing usable data storage. IPI-3 is designed for the software programmers who need to know only the generic characteristics of peripherals and who want to provide commands in a language not unlike that used by a high-level compiler.

On the other hand, *streaming tape* drives—both 1/2 and 1/4 in.—were conceived as backup devices for Winchester disks. In fact, the term streaming has become virtually synonymous with this backup function. Responding to the demands for standardization, the working group for *quarter-inch tape cartridge compatibility (QIC)* was formed. QIC spent its original efforts developing three standards: QIC-02, QIC-24 and QIC-36[16]. The three standards cover drive, media, and control compatibility and have helped establish *cartridge tape* as a viable product. QIC-02 specifies the host intelligent interface—the CPU's pipe to drive. QIC-24 defines the format in which data are recorded on the tape. QIC-36 defines the peripheral end of the interface. Functions at this end include tape control and phase locked loop design of the drive. The QIC group is continuing its study on new standards.

Terminal/Peripheral Interfaces

4.5 Interface Interference Problems and Solutions

Data transmission cable RFI/EMI problems were discussed in Chapter 3. In this section, interference problems associated with terminal and peripheral connections will be addressed. Representative examples which will be discussed include: noise suppression in computer grade connectors and peripheral actuators, conductive shield termination, use of optocouplers and EMP radiation detector circuitry.

4.5.1 Noise Suppression in Computer Grade Connectors and Peripheral Actuators

Transient suppression connectors, like the ITT Cannon shown in Fig. 4.9, can position a diode chip on each contact to shunt transient surges to ground.[21] This helps protect systems from damage. Clamping diodes can also be used at the mounting of connectors of ribbon cable on a PCB or backplane as depicted in Fig. 4.10.[22]

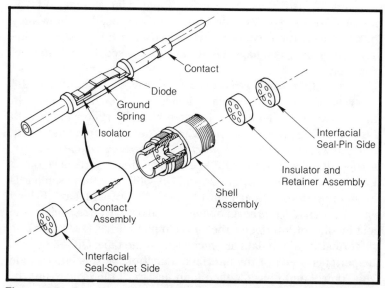

Figure 4.9—Transient Suppression by ITT-Cannon Connector Having a Diode Chip on Each Contact (from "Fewer Connections Translate into Fewer Failures," by B. Furlow, p. 105, Copyright by *Computer Design*, 1985, All Rights Reserved. Reprinted by Permission)

Computer-grade solenoids are used to move read/write heads in recording equipment. In many applications, rated solenoid response time can be greatly increased by applying a voltage spike of several times the rated operating voltage. The need to switch high current into an inductive load at a high speed and frequency causes *inductive kickback*, or extremely large voltage spikes which result from interrupting an inductive current. If not suppressed, these spikes can be many times the power supply voltage and emit very undesirable noise. Various noise suppression techniques help reduce inductive excursions (Fig. 4.11).[23] The simplest suppression scheme is the one using a diode across a solenoid (Fig. 4.11a). Two zener diodes connected back-to-back (Fig. 4.11b) can clamp transients of both polarities. Varistors that drastically lower their resistance upon application of a large transient have been used to short out a switched coil (Fig. 4.11c). A more traditional arc suppression circuit is shown in Fig. 4.11d, where a resistor in series with a capacitor connected across the coil will shunt coil current away from the inductance.

A closer look at the circuits of Fig. 4.11a to 4.11d reveals that all of these circuits have a common drawback—they extend *solenoid dropout* time by allowing current to circulate through the coil after it has been switched off. In applications that cannot tolerate even a small delay in coil current decay time, the circuit shown in Fig. 4.11e can be used. In this circuit, as soon as the switch interrupts the coil current, coil energy in the form of current is dumped into a capacitor. Then, when the switch contacts close, the arc suppression capacitor discharges slowly through the resistor.

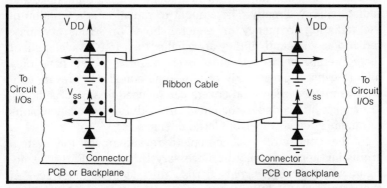

Figure 4.10—Clamping Diodes at Mounting Connectors of Ribbon Cable at I/O Devices (Reproduced from *EDN* by Permission of Cahners Publishing Co., Division of Read Holdings, Inc., and Author P. Mannone)

Terminal/Peripheral Interfaces

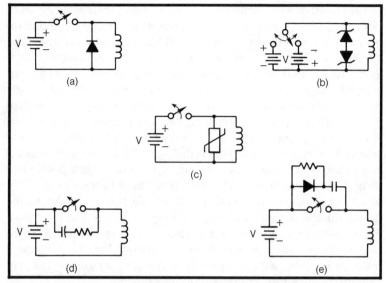

Figure 4.11—Noise Suppression Techniques in Peripheral Actuators with: (a) Common Diode, (b) Zener Diodes, (c) Varistor, (d) Shunt Current around Inductance (e) Circuit Extending Drop-Out Time (Adapted from Ref. 23)

4.5.2 Conductive Shield Termination at a Terminal or Peripheral Device

Sometimes a material's intrinsic shielding effectiveness is of less concern than is the leakage caused by shield discontinuities such as seams and holes. Holes can behave as slot antennae and radiate energy directly. The amount of radiation is a function of the radiating frequency. In general, holes or seams attenuate radiation significantly if they are smaller than 1/100 wavelength of the RF emission. Seams must overlap, or at worst, have only minute gaps. If particularly troublesome conditions prevail, conductive metal screens and covers can be used to seal large holes. They must be bonded to the enclosure with low impedance paths to minimize emissions from large openings.[24]

In the connector area, complete revamping of the system grounding approach may be necessary to achieve EMI-tight designs while getting signals across the electrical interface, as shown in Fig. 4.12. In this connection, conductive shield termination encloses conductors and extends enclosure shield to cable.

Optocouplers

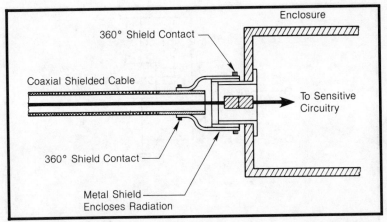

Figure 4.12—Conductive Shield Termination Encloses Conductors and Extends Enclosure Shield to Cable Shield

4.5.3 Use of Optocouplers as Interface Isolators[25,26]

In industrial environments, it is possible for very large transient common-mode voltages to exist between equipments. In cases such as these, an optocoupler can be used. The optocoupler TIL 107/108, shown in Fig. 4.13a, consists of an *infrared* emitter and receiver diode electrically isolated from each other but maintained in the same package. The emitter diode forward current is in the order of 15 mA to obtain a receiver current of 15 mA and to obtain a receiver current of 2 mA. A suitable buffer between the diode output and TTL input must be provided which, in the simplest form, would be a TTL Schmitt trigger. A maximum bit rate of approximately 60 kbps can be obtained. If higher bit rates are required, then the optocoupler should be followed by a wideband operational amplifier.

Figure 4.13b shows a high-speed optocoupler, the HCPL-2400, with a guaranteed data rate of 20 megabaud. This device offers very high common-mode transient immunity (typically 10,000 V/ μs). Applications include computer and telecommunications I/O interfaces, local area networks, data acquisition systems and switching-mode power supplies. An additional feature of this device is its tristate output which permits direct drive of data buses.

Terminal/Peripheral Interfaces

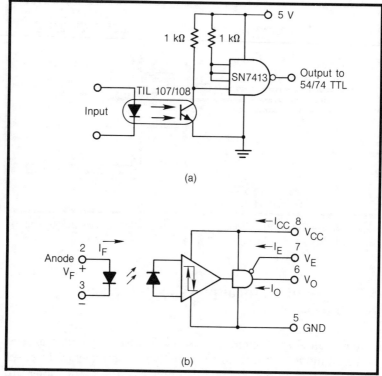

Figure 4.13—Optical Interfaces: (a) Low-Speed Optocoupler and (b) High Speed Optocoupler

4.5.4 EMP Radiation Detector Circuit for Terminal and Peripheral Protection[27]

Terminals and peripherals incorporated in critical command, communication, control and intelligence systems must continue to function after the initial radiation pulse dissipates.

Proper design methods can reduce the destructive effects of the *electromagnetic pulse (EMP)* and increase the chances of survival for an electronic equipment. The sequence of the detection and shutdown operations varies depending on the radiation dose rate, the speed of the system and how much disruption the circuit tolerates. Typical protection operations include the disabling of write operations for a hardened memory, the halting of CPU operations in a computer and its distribution of data and the control of power supplies. Some power supplies must operate throughout the period of exposure to nuclear radiation; consequently, their over- and undervoltage control circuits and overcurrent shutdown circuits must be temporarily shut off to avoid improper operation while radiation is present.

The operation of a complete detection and protection system begins as the control sequence starts when the *nuclear-event detector (NED)* senses a pulse of ionizing radiation. The detector's output signal triggers a monostable pulse generator, which generates a control pulse that has a preset duration. This pulse width must include time for the electronic components to recover from the burst of x-rays and gamma rays. Once back in their normal condition, the protected devices sense the end of the circumvention pulse, which signals the start of the recovery operations. The recovery operations reinitialize equipment or restore it to the state it was in when the detector sensed the radiation.

A 14-pin **hybrid nuclear-event detector (HNED)** module, the HSN-3000 chip, is shown in Fig. 4.14. This detector contains the four necessary functions: an ionization-radiation detector, a monostable pulse generator, a bistable latch and a built-in test circuit. A reversed-biased PIN diode (photodiode) within the chip acts as a fast-response detector which generates a photocurrent in response to ionizing radiation. The PIN diode's bias voltage (pin 8) can be externally adjusted to a maximum 20 V to improve photocurrent collection response and detection speed. Also, by varying the impedance between pin 9 and ground, it is possible to adjust the detector's dose-rate sensitivity.

Terminal/Peripheral Interfaces

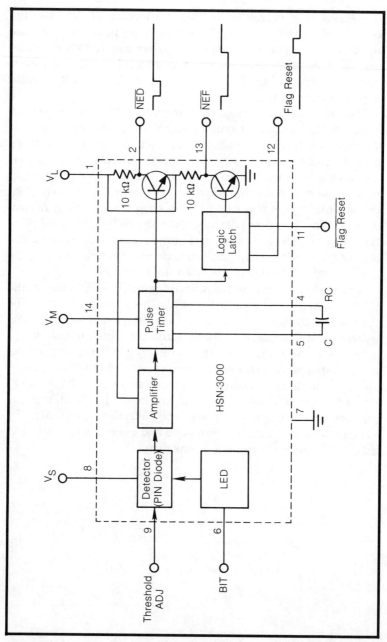

Figure 4.14—Hybrid Nuclear-Event-Detector Chip, HSN-3000 (Reproduced from *EDN* by Permission of Cahners Publishing Co., Division of Read Holdings, Inc., and Authors L. Longden and A. Trippe)

4.6 References

1. G. E. Huff and M. W. Green, *What Makes A Terminal Run*, Telecommunications, February 1984, p. 44.
2. J. Steeman, *RS232 Interface*, Elektor Electronics, September 1985, p. 76.
3. J.A. Sciulli and J.R. Peeler, *Error Analysis on Digital Communications Links*, Telecommunications, p. 51, March 1985.
4. M. Repko, *Modem Standards Aimed at Modern Transmission Schemes*, Computer Design, August 1, 1985, pp. 119–121.
5. N. Mokhoff, *AT&T's DMI Integrated Voice and Data Communications*, Computer Design, September 1, 1985, p. 42.
6. A. J. Weissberger, *Upgrade Data Communications With an RS-449 Interface*, EDN, February 17, 1982, p. 167.
7. E. E. Godsey, *"Big Applications for Small Terminals (With Small Prices)*, Wescon 1980 Professional Program, Anaheim, CA, September 16–18, 1980.
8. J. A. Titus, et al., *Interface Fundamentals: Hardware Encoded and Scanned Keyboards*, Computer Design, June 1979, p. 158.
9. R. A. Gilbert, *Operator Interface Design*, Hewlett-Packard Journal, July 1981, pp. 29–30.
10. W. E. Fletcher, *Moving the Man/Machine Interface Closer to the Man*, Wescon 1980 Professional Program, Anaheim, CA, September 16–18, 1980.
11. A. B. Caroll, *Three Types of Touch Technology Simplify Man/Machine Interface*, Computer Technology Review, Winter 1984, pp. 167–172.
12. R. Burgess, *RS-422 and Beyond*, Electronic Engineering, October 1981, p. 85.
13. W. J. Niewierski, *Providing CMOS Benefits to Peripheral Chips*, Computer Design, February 1983, p. 124.
14. D. Allan, *Varied Drive Interfaces Mystify Integrators*, Mini-Micro Systems, February 1985, p. 135.
15. M. Simmons, *How to Select Jumper Options for Floppy Disk Drives*, Mini-Micro Systems, February 1984, p. 253.
16. M. T. Makman, *Standards Pave the Way for Tape Cartridges*, Computer Design, May 1985, p. 133.
17. D. R. Ebenezer and N. Gordon *The Right Interface Can Improve Micro-To-Mini Data Transfers*, Computer Technology Review, Winter 1984, p. 100.

18. Advanced Micro Systems, Inc. *Am2900 Solutions for Systems, Peripheral Controllers*, Sunnyvale, CA, 1980.
19. C. Cory III, *Peripheral Controllers Provide the Versatility to Expand System Power, Computer Technology Review*, Fall/Winter 1981.
20. D. Hartig, *Peripheral Controllers Perform Tasks Once Handled by Host CPU's, Computer Technology Review*, Spring 1985, pp. 113–115.
21. B. Furlow, *Fewer Connections Translate Into Fewer Failures, Computer Design*, April 1985, p. 105.
22. P. Mannone, *Careful Design Methods Prevent CMOS Latch-Up, EDN*, January 26, 1984, p. 149.
23. D. Luckenbach, *Computer Grade Peripheral Acuators, Computer Design*, October 1981, p. 159.
24. E. VanderHeyden and J. H. Bogar, *Designing for Compliance With FCC EMI Regulations, Computer Design*, January 1982, p. 189.
25. B. Parsons, *TTL ICs Industrial Noise Environments, Texas Instruments: Power-Transistor and TTL Integrated-Circuit Applications*, McGraw Hill, NY 1977.
26. Components, *Twenty-Plus-Megaban Optocoupler for High-Speed Optical Isolation, h/p Measurement/Computation News*, September/October 1985, p. 7.
27. L. Longden and A. Trippe, *Hybrid Detector Protects Circuits from Radiation, EDN*, August 22, 1985, pp. 133–140.

General References

- R. Mercer, *Standards from a Common Carrier's Perspective*, Proceedings of Communication Networks Conference and Exhibition, Houston, TX, Jan. 12-15, 1981.
- M. Fox, *Concentrating on the Interfaces, Communications Engineering International*, November 1984, p. 33.
- M. Klimefelter, et al., *User-Defined Terminals Provide Flexibility for Changing Systems Needs, Computer Technology Review*, Summer 1983, pp. 79–81.
- C. Warren, *Computers and Peripherals, EDN*, July 16, 1982, p. 346.
- K. S. Padda, *Universal Peripheral Controller Frees CPU for High-Level Work, Electronic Design*, October 29, 1981, pp. 113–118.

References

- J. Olson, *Unformatted Controllers Solve Tape Drive Application Problems, Computer Technology Review*, Spring/Summer 1982, pp. 221–223.
- J. E. Jesson, *Smart Keyboards Help Eliminate Entry Errors, Computer Design*, October 1982, p. 137.
- R. A. Coilbert, *Operator Interface Design, Hewlett Packard Journal*, July 1981, pp. 29–30.
- C. J. Georgopoulos, *Interface Fundamentals in Microprocessor-Controlled Systems*, D. Reidel Publishing Company, Dordrecht, Holland, 1985.

Chapter 5
Data Bus Interfaces

This chapter deals with some functional aspects of buses that tie boards to data, control and power lines, and form the backbone of computer systems. In addition, the connection of various devices to a bus line as well as some annoying interference problems are discussed.

5.1 Introduction

Putting together a data processing or process-control system requires appropriate interface standards such as microcomputer and communications buses. A bus provides a standard physical and electrical environment so that a series of subsystems from many vendors may be integrated to form a unique control system. There are three main types of buses: a data bus that provides or accepts data, an address bus that points to storage locations and a control bus that provides or accepts control signals.

A bus line is shared by several devices that are directly or indirectly connected to it. Means for connecting such devices to the bus may include transformer coupling or direction connections of open collector devices, and differential line drivers and line receivers. In all of these cases the main concern of the designer is to reduce EMI and improve, if possible, the noise margin on the bus line. Interference problems can occur in many areas of a system, but the principal source is backplane signaling (board-to-board interconnections).

Data Bus Interfaces

5.2 The STD Bus

Of the bus systems available, the STD bus has become popular because its small card size provides cost-effective functionality. Cards are available which interface with thermocouples, pressure transducers, level sensors, cables, motors, valves, relays, solenoids and power controllers. Many are second-sourced, and new functions are being added on a regular basis.

5.2.1 Bus Description[1]

The *standard bus (STD bus)* is an industrial-quality bus for use in an 8-bit microcomputer-based system. The board modules that connect into the bus may be singular in function and can be placed in any slot, as shown in Fig. 5.1.

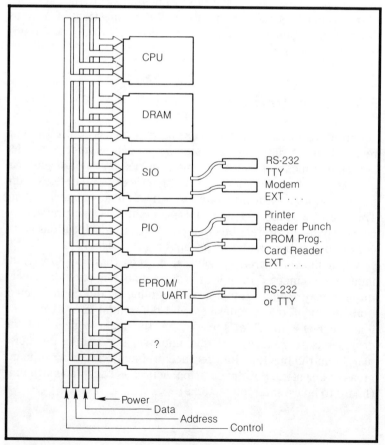

Figure 5.1—STD Bus System

The STD bus pinout is organized into four functional bus groups: power bus with pins 1 through 6 and 53 through 56, data bus with pins 7 through 14, address bus with pins 15 through 30 and control bus with pins 31 through 52.

The power bus can be up to five separate power supplies with two separate ground returns—one for digital circuitry and one for analog circuitry. The power pins are on the edges of the bus, and provide some EMI protection for the remaining signals between the power pins. Also, power is bused on both sides of the PC card. In addition to the main supply of +5 V, the bus provides two sources of -5 V and also +12 and -12 V for auxiliary circuits.

The data bus is an 8-bit, bidirectional, tri-state, high-level-active bus. All system cards are required to release control of the bus to a high-impedance state when not in use. For example, the processor card will release the data bus in response to a bus request input from an external system control.

The address bus is a 16-bit, tri-state, high-level active bus. The address bus provides 16 address lines for decoding either memory or I/O. Memory request and I/O request control lines differentiate between the two operations. The particular processor that is being used on the STD bus determines which lines are being used.

The signals in the *control bus* group control the operation of the bus. It is due to these signals that the flexibility of the STD bus is possible. Furthermore, the control signals can be subdivided into five junctional areas: memory and I/O control, peripheral timing control, clock and reset control, interrupt and bus control and serial priority chain.

5.2.2 Compatibility[1,2]

The purpose of any bus structure is to provide flexibility to the user in configuring a system to fit specific needs without compatibility problems. Each unique processor, however, on the STD bus requires specific timing characteristics to perform properly. Consequently, each CPU has a set of guidelines to assure that many, if not all products, work together. The STD bus specifications serve to define the usage of the bus lines. The timing specifications serve to guide each manufacturer as the designs products for the unique processor segment that he chooses.

The STD bus is designed for compatibility with industry standard TTL logic. Table 5.1 shows the specifications that apply for the STD bus. The STD bus supports the following processors:

Data Bus Interfaces

Z80™, 8080, 8085, 6809, 6800, 6512, 6502, NSC800 and more.

Table 5.1 STD Bus Logic Signal Characteristics

Parameter	Test Condition		Min	Max	Unit
VOH	V_{CC} = Min	IOH = −3mA	2.4	—	V
VOL	V_{CC} = Min	IOL = 24mA	—	0.5	V
VIH			2.0	—	V
VIL			—	0.8	V
Tr,Tf			4	100	NS

5.2.3 Real-Time Applications[3]

The STD bus *forte* is I/O intensive applications including real-time process and industrial control. A number of special I/O devices and STD bus peripherals have been developed by various manufacturers to suit the user's special needs. The peripheral timing control lines provided on the STD bus allow any 8-bit microprocessor to communicate with its one set of peripheral support chips. Most peripheral support chips are designed to operate with a specific processor. These signals permit the STD bus to handle any 8-bit microprocessor.

5.2.4 Intelligent Interfaces for STD Bus Systems[4]

Intelligent I/O interface boards are used in a system to enhance throughput over the STD bus and to more effectively use the computing power of the CPU. Each of these boards typically has its own microprocessor, random access memory (RAM), read only memory (ROM) and direct memory access (DMA) channel to the host memory and can perform peripheral control functions and data transfers to and from system peripherals independently of the host CPU.

Most intelligent boards for the STD bus offer the possibility of writing firmware to suit a system's peripheral. Often, firmware for communication between the host and the intelligent I/O board already exists. What is needed then is only to interface an application program to that firmware. In such a case, the main design task is to develop an application module that makes maximum use of the intelligent board's features to communicate with the devices linked to it.

Two typical intelligent I/O boards are the MDX-ISIO and MDX-I488 from Systems Technology. The MDX-ISIO has two serial communication channels with programmable band rates; the MDX-I488 acts as controller, talker or listener on a GPIB (see following subsection) instrumentation network. Systems using either ISIO or I488 intelligent boards outperform any single-CPU system, principally, because many of the operations which would normally be performed by the central CPU can be offloaded to the intelligent board and run independently.

5.3 General Purpose Interface Bus (GPIB) or IEEE Std 488-1978

General-purpose interface bus (GPIB) is a method used throughout the world for linking instruments and computers. Having first been considered some 20 years ago, GPIB is still the foremost standard for connecting instruments to computers.

5.3.1 Definitions and Equivalent GPIB Standards

GPIB is defined internationally by the IEC625-1 standard and in the United States by ANSI MCI1.1 and IEEE Std 488-1978. The bus is also covered by a British standard, BS 6146. Sometimes it is called HP-IB because its initial design started with Hewlett Packard back in 1965.

The GPIB (IEEE 488) has a party line structure where all devices on the bus are connected in parallel (Fig. 5.2). Therefore, the cable's role is limited to that of interconnecting all devices in parallel, whereby any one device may transfer data to one or more other participating devices.

Table 5.2 gives the description of the bus lines. The interface functions for the bus are defined as follows:[5]

1. *The Source Handshake function (SH)* controls the transmission of data bytes over the data bus using the handshake bus.

Data Bus Interfaces

2. *Acceptor Handshake function (AH)* oversees proper reception of data bytes using the handshake bus.
3. *Talker function (T)* enables a device to transmit data bytes over the data bus. It should be noted that an extended listener is also defined and designated (TE).
4. *Listener function* allows a device to receive data bytes over the data bus. Again, an extended listener is also defined and designated (LE).

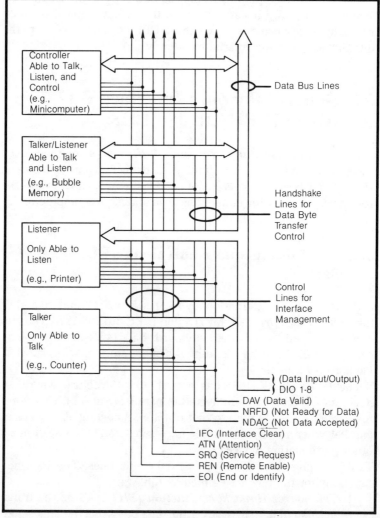

Figure 5.2—GPIB/IEEE 488 Active Signal Lines for Multiple Devices

GPIB Standards

5. *Service Request function (SR)* allows a device to manipulate the service request line in the management bus and thereby request service asynchronously of the controller.
6. *Remote/Local function (RL)* allows a device to obtain programming information from one of two sources locally via the panel or remotely via the bus.
7. *Parallel Poll function (PP)* makes a parallel poll of devices possible; that is, it enables the controller to interrogate all of its dependent devices simultaneously in order to determine which of them requested service.
8. *Device Clear function (DC)* makes it possible for the controller to clear a device or a group of devices to a known state.
9. *Device Trigger function (DT)* allows the controller to trigger a device or a group of devices to commence operation. There are often easier ways of achieving this end because trigger commands are often included as part of peripheral programming strings.
10. *Controller function (C)* manages the operation of the bus system, primarily by designating which devices are to send and receive data. It may also command specific actions within other devices.

Table 5.2 The IEEE—488 lines

Designation	Description
DIO_1—DIO_8	*Data Input/Output:* Eight data transfer lines: also called the data bus.
ATN	*Attention:* Issued only by the controller, to gain the attention of bus devices before beginning a handshake sequence and to denote address or control information on the data bus.
DAV	*Data Valid:* Issued by a talker to notify the listener(s) that data has been placed on the DIO lines.
EOI	*End or Identify:* Issued by a talker to notify the listener(s) that the data byte currently on the DIO lines is the last one. The controller issues is together with ATN to initiate a parallel poll sequence.
IFC	*Interface Clear:* Issued only by the controller to bring all active bus devices to a known state.
NDAC	*Not Data Accepted:* Issued by a listener while fetching data from the DIO lines.
NRFD	*Not Ready for Data:* Issued by all listeners and released by each listener as it becomes ready to accept data.
REN	*Remote Enable:* Grounded to the maintain control over the system.
SRQ	*Service Request:* Issued by any device needing service from the controller.

5.3.2 Incompatibilities and Limitations

With respect to operating data rates, the IEEE 488 specifications state that using 48 mA open collector drivers, a data rate of 250 kBps can be obtained for distances up to 20 m by using an equivalent standard load for each 2 m of cable within the system. For higher transfer rate requirements, up to 500 kBps may be achieved in the same system by substituting 48 mA tri-state drivers in place of open collectors. For those applications requiring even higher speed, it is possible to achieve data rates up to 1 MBps by using the following guidelines:
1. All devices must use tri-state drivers.
2. Capacitance of each lead of the interface must be less than 50 pF per device.
3. All instruments in the system must be powered on.
4. Cable length should be limited to a maximum of 15 m with at least one equivalent load for each meter cable.
5. All devices expected to operate at this high transfer rate must be capable of handshaking data at the data rate desired.

When considering compatibility, simply saying that the instrument is compliant to the IEEE 488 is not sufficient for total evaluation of how the instrument will perform in an automated system. It only means that the instrument has an interface implementing some of the IEEE 488 capability. Because it takes advantage of a large selection of programmable instruments and stimuli, the IEEE 488 bus has proven to be a good method for interconnecting instruments and computers. Large systems, however, are restricted by two major bus limitations.[6,7] Driver load capacity limits the system to 14 devices. This is far less than the primary address limit of 31 devices or, if secondary addresses are used, 961 devices. Cable length limits the controller device distance to 2 m per device or 20 m total, whichever is less. This imposes transmission problems on systems spread out in a room or on systems that require remote measurements. Although the IEEE 488 standard specifies electrical shielding to provide more noise immunity than is available with RS-232 and many other bus structures, noise caused by ground loops can and often does interfere with proper equipment operation, particularly as most IEEE 488 implementations employ 5 V bipolar logic.

The IEEE 488 bus is flexible and more than adequate for interconnecting several moderately complex and intelligent test instruments with the processor. It is not suitable for general purpose I/O interconnection from a microprocessor to relatively modest user-generated printed circuit boards, such as is normally required for automatic test equipment (ATE).

5.3.3 Using Expanders to Extend the IEEE Std 488 Bus Capabilities

When using IEEE 488 bus interfaces, the system problems are associated with the need for more devices, longer cable length and remote bus. Using bus extenders that are available from various manufacturers, it is possible to effectively solve these problems.

For example, a *bus amplifier* can be used to solve the electrical loading problem. Such an amplifier has been incorporated in the system of Fig. 5.3 termed *bus expander*.[6] Specifically, this device is the ICS Model 4830 Isolator/Expander which replaces the last device on the original bus and drives up to 14 additional devices on an isolated bus. The major advantages of the 4830 are its moderate cost, its isolation of the added devices from the original bus system and the fact that it causes almost no bus-speed degradation. Similarly, it has virtually no effect on the overall system response, as it adds less than 200 ns of propagation delay to the bus signals. If it becomes necessary to expand the system beyond 27 devices, obviously more bus expanders must be added. The 4830 does not use a system address and, insofar as the bus controller is concerned, is fully transparent.

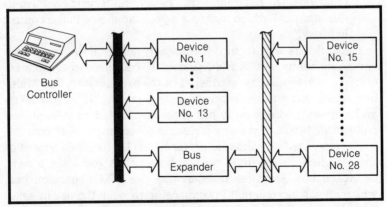

Figure 5.3—Expanding GPIB's Load Capabilities Via Bus Amplifier

5.4 Multibus System Bus (IEEE P796)

Multibus architecture in a multiprocessor system allows each processor to work asynchronously. Therefore, a fast microprocessor operates at its own speed regardless of the speed of the slowest microprocessor. This technique tolerates duty cycle and phase shift variations and offers hardware modularity. When new system functions are desired, additional microprocessors can be integrated without impacting existing task partitioning.

5.4.1 Definitions and Evolution of Multibus Systems[8-10]

Some definitions are given first concerning the use of Multibus systems in multicomputing and multiprocessing equipment design.
1. *Multicomputing* is a top-down design philosophy which is based on a functional partitioning of the solution of a particular problem into separate, well-defined, functional modules.
2. *Multiprocessing* is a system employing two or more processing units which share resources and interact under integrated control.
3. The *Multibus System Bus*, introduced by Intel in 1976, has become the industry standard commercial-quality microcomputer bus. Over 100 industrial peripheral boards conform to this standard, which makes it a popular choice in the control industry.

The Multibus architecture was initially designed to handle 8-bit communications, but introduction of 16-bit microprocessors required a corresponding expansion of the architecture. The Multibus architecture today supports both 8-bit and 16-bit transfers, and memory addressability has expanded to 24 bits. A Multibus-compatible board physically measures 17.14 cm × 30.48 cm.

The increased demands for performance led to enhancement of the Multibus design. Intel accomplished this by adding a new, dedicated bus. The first addition was the iSBX expansion bus, which allows incremental I/O expansion through the use of small iSBX Multimodule expansion boards. Later, the Multibus I/O bus was added. This allowed Multibus-based systems to accommodate I/O transfers as fast as 8 MBps at distances up to 15 m. This offers an alternative to the slower speed, but somewhat longer distance, of the IEEE 488 bus. The Multichannel bus can support as many as 16 devices and provide 16 MB of memory or address register space per device.[8]

Still later came the iLBX *local bus extension*, which allowed the processor board to access up to 16 MB of off-board memory, but with performance compatible to that of on-board memory.

5.4.2 Multibus II Products

Six board-level products represent the first *Multibus II* boards that Intel Corporation has introduced. Multibus II uses six buses: a parallel system bus, a local bus extension, a serial system bus, an expansion bus, a multichannel DMA I/O bus and a bitbus serial I/O expansion bus. The first of these buses, the *parallel system bus (iPSB)*, is processor-independent. It provides data movement and interprocessor communications and has a burst-transfer capability of 40 MBps (maximum sustained bandwidth). A 1-bit-wide *serial system bus (iSSB)* offers a software-compatible alternative to the parallel system bus for multiprocessing. The data transfer rate is 2 MBps.

A *local bus extension (iLBX)* provides arbitration-free local memory expansion up to 64 MB. The extension supports 8-, 16- or 32-bit microprocessors with up to five memory-expansion boards each. An *expansion bus (iSBX)* functions with small multimodule expansion boards to provide greater system utility without adding another full board. A multichannel DMA I/O bus provides 8 MBps data transfer between a single-board computer and as many as 16 mass storage devices or distributed peripherals. It provides 16 MB of memory or register address space for each device. The *bitbus* serial I/O expansion bus serves as a microcontroller interconnection for both intrasystem and intersystem communications. It functions with a standard synchronous data-link-control protocol and RS-485-compatible format. Operating speeds can vary within the 62.5 Kbps to 2.4 Mbps range.[11]

A tool has been developed lately for interfacing Multibus II and I devices. The new tool is the iSBC LNK/001 board of *Intel* that serves as a translator between developmental Multibus II systems and existing Multibus I products, enabling designers to make use of capabilities of both versions. The link board has a Multibus I form factor, resides on the Multibus I board and connects to Multibus II boards via the P2 connector. The link board allows Multibus I boards to act as slaves to the Multibus II boards.[12]

5.4.3 Signal Conditioning of Multibus Compatible Analog I/O Modules

Multibus compatible analog I/O modules provide measurement and control capability for a range of data acquisition and voltage-actuated applications. Analog I/O boards for the Multibus can amplify a low-level signal but provide only minimal filtering, input protection and isolation. Input protection in an analog input board using CMOS multiplexers is limited to about 35 V differential channel-to-channel. Without further isolation, any electrical fault of the input wiring will be coupled into the I/O board and onto the system backplane, leading to possible system damage. *External signal conditioners* overcome these limitations and also provide additional signal conditioning, such as excitation for bridge transducers and cold junction compensation for thermocouples (Fig. 5.4).

For original equipment manufacturers (OEMs), system houses and end-users, the most popular external conditioners are *signal conditioning manifolds*.[13] The signal conditioning manifold typically provides 16 channels of both input and output signal conditioning. An individual signal conditioning module connects to each channel on the system backplane. In certain cases, a ribbon cable can be used with a system for direct connection from the manifold's backplane to the edge connector of an analog input or output board.

The backplane also contains the power supplies to operate the signal conditioning modules connected to the backplane and screw-terminals for connection to the sensors. Manifold input modules can accept a high- or low-level *dc* or *ac* input and provide a $+10$ V or $0-10$ V output to the input connector of the analog I/O board. Some manifolds also produce a simultaneous 4–20 mA or 0–20 mA output for remote transmission of sensor data to external devices. A manifold output module takes a ± 10 V or $0-10$ V signal from an analog output board and converts this to a 0–20 mA or 4–20 mA output. The primary differences between the

Signal Conditioning

signal conditioning manifolds are the specifications and range of input or output modules. At best, a manifold system should offer a module to condition each of the various signals expected to be encountered in the application of interest.

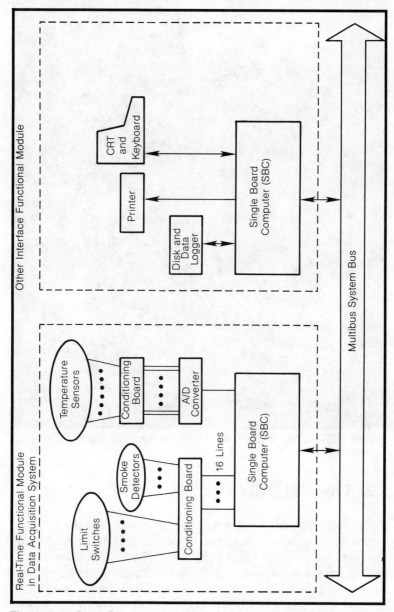

Figure 5.4—Block Diagram of Multibus System Bus with External Signal Conditioning

Data Bus Interfaces

Figure 5.5 shows a Multibus board-level product, the RTI-711 of Analog Devices, which provides 16 single-ended, or 8 differential, analog inputs with 12-bit resolution.

Figure 5.5—A Multibus® Board-Level Product (Courtesy of Analog Devices)

5.5 The VME Bus

The *VME bus* and its extension bus structures form the basis of a total *VME system* architecture which simplifies integration of complete systems from high-performance 8-, 16- and 32-bit board level system components. Mostek, Motorola and Signetics/Philips have agreed on detailed technical specifications for the new VMX bus and VMS bus, for which documentation has been developed and made available.[14]

5.5.1 VME System Features

Since its introduction in 1983, VME bus has grown, and today there are more than 90 vendors selling over 300 VME bus products.[14] This fast growth rate led the IEEE to establish a committee (IEEE P1014) to create an industry standard based on VME bus.

The VME system architecture is primarily a hardware and protocol specification used to interconnect the computing, storage and communication elements of a control system. VME is best suited to applications where more processing power is required, such as the central controller in a distributed process control, work station, main robot controller and large dedicated machine controllers. The features that make VME suitable for these applications are:[15]

1. The VME system supports 8-bit, 16-bit, and 32-bit processors and data transfer.
2. The interconnection is specified for high performance (non-multiplexed, high speed).
3. VME contains the elements to easily support growth through multiple processors and/or wider processors.
4. The form factor is Eurocard configuration with pin-in-socket connectors, highly regarded for industrial applications.
5. VME provides utilities that support self-test and diagnostics.

The VME bus provides the primary system data transfer mechanism in a VME system. A parallel, nonmultiplexed data transfer bus with byte (8-bit) and word (16-bit) data transfers, and with short I/O (16-bit) and standard (24-bit) addressing spaces is implemented on the primary connector P1. Expansion to full 32-bit address space and data transfer (long word) is provided by the P2 connector. All control, VMS bus, priority interrupt, bus arbitration and utility signals are located on the primary connector so that a full-function, non-expanded (16-bit data, 24-bit address) VME bus interface can be supported on the single high card. The second 96-pin connector has 32 lines reserved for VME bus expansion (16 data, 8 address, plus power), and the remaining 64 lines are used either for user I/O, or for a VMX bus implementation. Table 5.3 summarizes VME bus characteristics.

Data Bus Interfaces

Table 5.3 Summary of VME Bus Characteristics*

Bus type	Asynchronous. Non-multiplexed
Address width	24 Standard
	32 Expanded
Data width	8, 16, 32 (expanded)
Board size	Double or single high eurocard
Connector type	Pin & socket
Primary/secondary connector pins	96/96
Supply	+5 ± 12
Voltages (V)	+5 Standby
Interrupt levels	Seven
Arbitration levels	Four
Multiprocessor?	Yes
Error signals	AC fail. System fail, bus error
Special cycles	Read/modify/write, block transfer. Access privilege levels
Expandable modes for future	Yes (address modifiers)

*Reprinted by permission from Motorola Systems Design News © Motorola 1985.

5.5.2 VME System Architecture

The VME system architecture is based on three bus structures: the VME bus, the VMX bus and the VMS bus (Fig. 5.6).[15] The heart of the VME structure is the VME bus interconnect standard which provides the basic parallel data transfer bus between system components. VMX bus facilitates expansion of a local processor's memory by extending its local high performance bus. The VMS bus is different from other buses because it is a self-arbitrating serial bus using two signal lines to pass messages among functional modules within a backplane or in separate card racks.

The VME bus has a master/slave asynchronous data transfer format which supports data transfer rates in excess of 20 MBps in the expanded 32-bit mode. The asynchronous protocol allows mixing devices of varying speed without slowing the bus to the lowest system device. The three basic transfer cycle types are read, write and nondivisible cycles.

VME and Multibus II

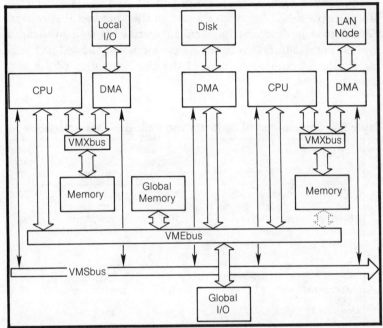

Figure 5.6—VME system architecture is configured around three bus structures; the VMEbus, VMX bus, and the VMSbus (Reprinted by permisiion of *Motorola Systems Design News*, © 1985, Motorola.

5.5.3 VME Bus and Multibus II Comparison[16]

VME bus and Multibus II use multiple buses serving dedicated functions to ensure the high throughput afforded by 32-bit-wide data paths. Multibus II, an upgrade of the 16-bit Multibus I, includes five buses in its specification; VME bus has four buses. But there are some functional similarities between them. Both have a parallel system bus (iPSB, VME bus) for interboard communications and data movement, a local high-speed bus extension to memory (iLBX, VMX bus) and a lower cost serial alternative to the parallel system bus (iSSB, VMS bus).

On the other hand, Multibus II carries over two buses unchanged from Multibus I: the iSBX I/O expansion bus, permitting the addition of multi-modules to processor and other system boards, and the multichannel I/O bus for direct-memory-access block transfer to intelligent I/O devices. VME bus' I/O channel,

Data Bus Interfaces

which serves local I/O expansion up to 3.66 m, is similar to Multibus II's iSBX. Although no part of the Multibus II specification, *Bitbus* is designed for local industrial I/O to 3 m using a synchronous data link control-like protocol and twisted-pair wiring. Table 5.4 compares some of the characteristics of the two types of buses.

Table 5.4 Comparison of Multibus and VME Bus Characteristics

	Function	Maximum data rate (bytes per sec.)	Type	Data width (bits)
iPSB	parallel system bus	40M	synchronous, multiplexed	32
iLBX II	local high-speed memory extension	48M	synchronous, nonmultiplexed	32
iSBx	local I/O expansion	10M	asynchronous, nonmultiplexed	16
Multichannel I/O	rer e DMA	8M	asynchronous, multiplexed	16
iSSB	lowe ost serial sys m bus	2M	CSMA/CD protocol	1
Bitbus	local industrial I/O	2.4M	protocol similar to SDLC	16
VMEbus	parallel system bus	40M	asynchronous	32
VMXbus	local high-speed memory and I/O extension	80M	asynchronous	32
I/O Channel	local I/O expansion	2M	asynchronous	8
VM⌒bus	lower cost serial system bus	3M	token-passing	1

5.5.4 IBM PC/AT VME Adaptor

The IBM PC/AT VME Adaptor interconnects an IBM PC/AT to a VMEbus system. The PC/AT can be used as a bus master processor on the VMEbus or as a coprocessor in a multiprocessor application. Memory windowing maps up to 16 MB of VMEbus addresses onto the PC/AT address range. This enables the PC/AT to perform random access memory reads and writes on the

VMEbus. Memory, I/O and all other VME address modifier modes are supported. A dual port RAM option (to 1 MB) plugs into the VME card to serve as a common memory to the PC/AT and VMEbus. With the Dual Port Ram option, access to the memory by one bus does not use bandwidth from the other bus.

The two printed circuit cards are connected together with a round EMI-shielded cable. Transfers are parity checked on address, control and data. Both word and byte operations are supported.

Figure 5.7—IBM PC/AT VME Adaptor (Courtesy of Bit 3 Computer Corporation)

5.6 Other Types of Data Buses

This section presents some additional versions of bus structures which respond to various degrees of complexity, and the requirement for compatibility with past and future 8-, 16- and 32-bit systems.

5.6.1 Brief Descriptions and Special Features

There are more boards available for the *S-100* bus than any other, but its ambiguously defined contacts can cause problems and a lot of unplanned rewiring. The ±8 and ±16 V inputs are not standard, and every board requires its own voltage regulation. There are several spare lines though, which is an advantage for custom designed peripherals.[17]

The *VERSABUS* was conceived to be a high-performance 16-bit computer bus with 32-bit expansion capabilities. VERSABUS is supported by available cost effective hardware such as the 68000 family octal bus drivers and receivers.[18,19]

The *BITBUS* is a standard serial connection for microcontrollers and provides the requisite features of a distributed control bus. Low cost hardware and software implementation, high reliability with consistent error detection and recovery, ease of development and use and high performance with speed/distance options are specific architectural objectives for the targeted distributed control environment.[20]

The *MIL-STD-1553B* bus has become well-established as a data bus system standard for military aircraft (and its LSI implementation has aroused considerable interest elsewhere). It has been adopted within the United Kingdom as Defense Standard 0018 and in NATO as STANAG 3838. However, while 1553 has revolutionized the design of avionic systems, it has presented equipment manufacturers with the problem of testing these systems.[21]

The *NuBus* is a synchronous bus; all transitions and signal sampling are synchronized to a central system clock. However, it has many of the features of an asynchronous bus; transactions may be a variable number of clock periods long. This provides the adaptability of an asynchronous bus with the design simplicity of a synchronous bus.[22]

The IEEE P896 *Futurebus* is an advanced, high performance, backplane bus designed before in order to support multiple generations of multiprocessor architectures over a long design lifetime. P896 is an asynchronous, 32-bit address/data multiplexed bus, with every aspect of its interface fully distributed and requiring no central control of any kind.[23]

The *STEbus* Eurocard standardization allows cards designed around it to function in Eurocard-based dual-bus systems. In such a system, a high-end 32-bit bus, such as the VME bus, uses STE cards as low-cost I/O. An STE board interfaces neatly with the two-connector VME scheme because the STE scheme uses only the outside two rows of the P2 DIN connector.[24]

The long history of Digital Equipment Corp. has brought widespread acceptance of the *Q-bus*. A large number of hardware and software developers continue to support this bus, using mainframe power at a reasonable cost.

5.6.2 Some Comparisons

A large part of the electronics industry's growth in recent years must be attributed to the ability to configure specialized systems which use the processing power of each new microprocessor. It is the increasing degree of complexity, the requirement for compatibility with past and future versions of a particular system and the need to communicate with other systems that has led to adoption of the numerous 8-, 16- and 32-bit bus structures.

Tables 5.5[26] and 5.6[27] compare the main parameters of some of the leading buses that have been already described. These tables are very useful to system designers, since they contain information regarding the capabilities and limitations of various bus types and make their choice easier.

It must be noted that besides the buses that were discussed in this chapter, there are several other types available on the market, while some others are in the development stage. However, a more extensive treatment of bus types here would be beyond the scope of this book.

Data Bus Interfaces

Table 5.5 Bus Comparison Chart (1)

Bus Type	STD Bus	Multibus	S100	HPIB	Q Bus	Versabus	VME
Developer	Pro-log/mostek	INTEL	IMSAI	Hewlett-Packard	DEC	Motorola	Motorola/others
Year introduced	Approx 1977	Approx 1976	Approx 1975	Approx 1973	Approx 1978	Approx 1979	Approx 1982
Public specification	Yes	Yes	Yes	Yes	Yes	Yes	Yes
Controlling body	Mfgr group IEEE prop	IEEE 796	IEEE 696	IEEE 400		IEEE 970 (proposed)	
Primary application	Computer I/O backplane	Computer backplane	Computer backplane	Instrument interconnect	Computer I/O backplane	Computer backplane	Computer I/O backplane
Width Address (bits) data	16 8	24 8 or 16	16, 24 8 or 16	4 8	16, 22 8 or 16	16, 24, 32 8, 16, 32	16, 24, 92 0, 16, 32
Protocols	Master/slave	Multiprocessor	Multiprocessor	Multi master/slave	Master/slave	Multiprocessor	Multiprocessor
Card size Area (")	4.5 × 6.6 29	6.76 × 12 81	6.1 × 10 61	None	6.26 × 8.9 47	9.26 × 14.6 134	6.3 × 9.2 68
Processors supported	0085, 0088 Z00, 6009	88XX, Z80 60XX, 60000	80XX, Z80 60XX, 68000	I/O only	LSI-II	60000	60000, 80186

160

Some Comparisons

Number of functions*	1000–1200	800–1000	600–800	400–600	100–150	100–300	160–300
Number of vendors*	120+	160+	100+	100+	20+	10+	25+
Cost/function	Lowest	Medium	Low medium	High—Complex functions only	High	Med high	Med high
Main user types	Industrial/laboratory	Computer based applications	Hobbiest/general purpose	Instrumentation/systems	General		
Remarks	Multiprocessor specd—not implemented	Largle—multi function boards	On board power regulation	Limited ● of I/O devices	Expensive	Motorola version of multibus	Newest of the bus systems
	Rugged—for industrial use	Complex for simple systems	Heat causes reliability problems	Requires intelligence at both ends of I/O	Minicomputer based systems	Big boards	Big in Europe fast growth projected

*Latest estimates
(1) **Reprinted from Ref. 25 by kind permission of the author R. L. Mack.**

Data Bus Interfaces

Table 5.6 Parameters of Some of the Leading Buses*

	STDbus	STEbus	S100bus	Multibus I	Versabus	VMEbus	Multibus II	Nubus	Futurebus
IEEE standard or project number	P961	P1000	P696	796		P1014	PXXX	PXXX	P896
Data width (bits)	8	8	8/16	8/16	8/16/32	8/16/32	8/16/32	8/16/32	8/16/32
Maximum address (bytes)	64K	1M	16M	16M	4G	4G	4G	4G	4G
Protocol	nonmux sync	nonmux async	nonmux sync	nonmux async	nonmux async	nonmux async	nonmux sync	mux sync	mux async
Number of processors	1+1+	1+2	1+15	1+3	1+4	1+4+	20	16	32
Arbitration	chain	parallel	parallel	chain	parallel + chain	parallel + chain	parallel distributed	parallel distributed	parallel distributed
Size (mm)	114.3 × 165.1	standard single/double Eurocard	152.4 × 254.0	171.5 × 304.8	368.3 × 234.9	standard/extended single/double Eurocard	double Eurocard	hyper extended triple Eurocard	hyper extended triple Eurocard
Connectors	56, edge	64 (of 96) DIN	100, edge	86+60, edge	140+20, edge	96+32 (of 96) DIN	96 DIN	96 DIN	96 DIN
Processors (type)	Z80, 8080/85 6502, 6800/9	Any 8-bit	Z80, 8080/85 80186, 68000, Z8000, etc.	Z80, 8080/85 8086, 68000, Z8000, etc.	68000, 68020 others possible	68000, 68020 others possible	Any (16-, 32-bit)	Any (16-, 32-bit)	Any (16-, 32-bit)
Bus drivers	TTL	TTL	TTL	TTL	TTL	TTL	TTL	TTL	Special

*Courtesy of Advantec Systems.

5.7 Device Connections and Interference Problems in Bus Organized Systems

A bus line is shared by several devices that are directly or indirectly connected to it. Methods of connecting such devices to the bus may include transformer coupling (or direct connections of open collector devices) and differential line drivers and line receivers. In all of these cases, the main concern of the designer is to reduce EMI and improve, if possible, the noise margin on the bus line. Interference problems can occur in many areas of a system, but the principal source is backplane signaling (board-to-board interconnections).

5.7.1 Bus Transformer Coupling in MIL-STD-1553B[21,28]

One serial bus structure is defined in MIL-STD-1553B, which establishes the requirements for serial, digital, command/response and time-division multiplexing techniques on aircraft. The bus allows for the transmission of information among several signal services interconnected by a single twisted shielded-pair of wires. The bus is defined as a transformer-coupled, fault-isolated, transmission line with data appearing differentially on the signal-carrying wires.

In Fig. 5.8a, the coupling transformer connects through a stub less than 6 m long at one end and to an N:1 isolated transformer at the other end. For connections where the stub length can be less that 0.3 m, a direct-coupled connection (Fig. 5.8b) can be substituted. Direct coupling requires only a 1:1.4 coupling transformer and two 55 Ω fault-isolation resistors for interfacing the data bus to the transceiver. Transformer-coupled stubs are more desirable since they provide more protection to the transceiver electronics should a short circuit occur at the stub.

Since the MIL-STD-1553B bus must operate in a military environment, it is largely immune to electrical noise interference. Significant levels of EMI from ground loops, jamming signals and spurious emissions have less effect on 1553 systems than on parallel buses. These characteristics can be advantageous in many high-noise industrial applications, especially if optical isolators are used instead of transformers. As a further improvement, fiber optic cable with superior EMI and isolation characteristics may replace the twisted pair in the next generation of data buses.

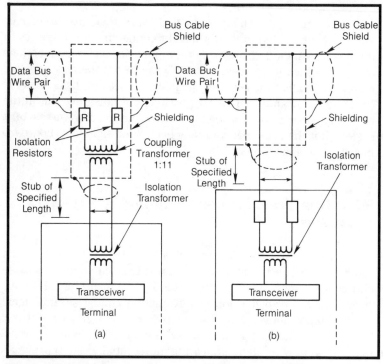

Figure 5.8—Data Bus Interface Using (a) Transformer Coupling and (b) Direct Coupling

5.7.2 TTL-System Bus Line with Combined Active/Passive Terminations[29,30]

The major design problem in interfaces with bus lines is the multiplexing of solid state devices: that is, the direct connection of several such devices onto a single bus line to perform or function which is very important. Additionally, an ideal pair of driving/receiving circuits should present as small an ac and dc load as possible to the bus line. A wired OR is not possible in conventional TTL circuits or their equivalents because they provide only two states at their output, namely "0" and "1," using both active *pull-up* and *active pull-down* stages.

TTL Bus with Terminations

In saturated type logic, as in the TTL family, dc noise margins are defined in terms of the voltage levels at the output of the driving gate and those input levels required by the driven gate to recognize, without ambiguity, a "1" or "0" input level. The difference between the output voltage levels guaranteed for the driving gate, the input voltage levels guaranteed for the driving gate and the input voltages required for the driven gate defines the low-state noise margin (NM_L) and the high state noise margin (NM_H), i.e.,

$$NM_L = (V_{IL})_{max} - (V_{OL})_{max} \quad (5.1a)$$
$$NM_H = (V_{OH})_{min} - (V_{IH})_{min} \quad (5.1b)$$

The difficulty here is with the low-state noise margin ($NM_L \approx$ 400 mV guaranteed max). Any reduction in this maximum "0" level will constitute an additional margin improvement. When a transistor is connected in the inverted mode of operation, the emitter current is related to the base current by the inverted common-emitter current gain β_I as follows (see also Fig. 5.9):

$$I_E = \beta_I I_B \quad (5.2)$$

Emitter current is into the emitter and collector current is out of the collector (npn transistor). This allows the possibility of using active line termination in addition to ground return termination as shown in Fig. 5.10.

When a transistor is used in this manner, very low emitter-to-collector saturation voltage can be realized. Typically, this voltage is only a few millivolts. The inverted current gain of a transistor is usually much less than its normal current gain, generally by a

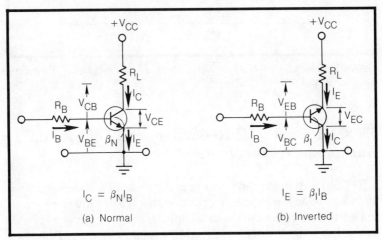

Figure 5.9—Transistor Operating Modes

Data Bus Interfaces

factor of ten or more. The difference between the normal-mode and inverted-mode "0" voltage levels gives the magnitude of logical "0" dc noise margin improvement or NMI, i.e.,

$$\text{NMI} = V_{CE \text{ sat (norm)}} - V_{EC \text{ (invert)}} \quad (5.3)$$

Tests showed that by proper transistor selection and biasing levels, an inverted-mode saturation voltage of 0.04 V or less can be maintained for 20 loading gates; so then, Eq. (5.3) yields NMI = 0.36 V for the logical "0" state of the bus line. It should be noted that during the steady-state "1" condition, the terminating transistors go out of saturation, thus allowing the bus line to maintain its high state.

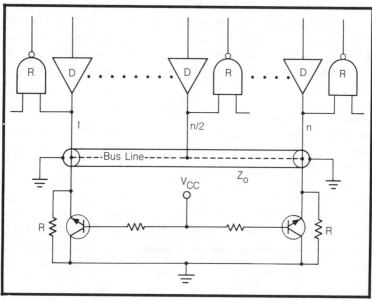

Figure 5.10—Bus-Line System with Combined Passive/Active Termination (Adapted from Ref. 30)

5.7.3 RFI and EMI Reduction with Differential Backplane Transceivers[31,32]

RFI and EMI noise can occur in many areas of a system, but the main source is backplane. To combat this type of interference, a transceiver with the means to comply to FCC regulations should be used. A suitable device is the AMD's Am26LS38 *Quad Differen-*

RFI/EMI Reduction

tial Backplane Transceiver. It is designed to integrate Schottky TTL performance, high noise immunity and wired logic capability into low cost differential backplane structure.

The Am26LS38's differential signaling provides two forms that contribute to cancellation. First, radiated signal caused by the dv/dt in each signal load is opposed by its equal and opposite complement, thus cancelling the RFI. Second, terminating the cable pair correctly line-to-line in its characteristic impedance assures that current flowing in one signal lead is opposed equally by its mating pair, thus minimizing EMI. Proper termination is shown in Fig. 5.11. The Am26LS38's driver-receiver pairs are capable of operating in a balanced differential mode without distortion.

The receiver is designed with a low (±50 mV) threshold that combines with its driver output of greater than 500 mV to provide systems attenuation tolerance and margin for the cable. A 25 mV (typical) hysteresis minimizes the switching sensitivity caused by lines with slow transition times. The device can operate up to 1.5 V of common-mode voltage, thus eliminating the need for absolute ground references. To reduce radiated harmonic disturbances, the Am26LS38 features symmetrically controlled rise/fall times.

In shared-bus, digital-balanced line transmission the device performs especially well for interelement backplane communication within a processor complex. In such applications, the bus line is bidirectional and party lined, but the subsystem internal buses are unidirectional. While network dimensions vary with the types

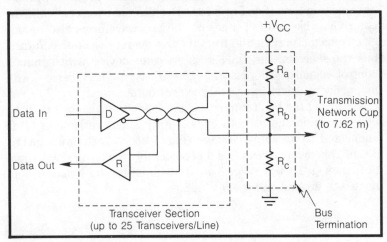

Figure 5.11—Termination of AM26L538 Devices in the Backplane

Data Bus Interfaces

and sizes of cabling used, a typical network of up to 7.62 m (and up to 25 transceivers per line) can be configured with data rates in excess of 10 MHz.[31]

Another example of noise immunity and EMI reduction in a backplane is the Futurebus. The Futurebus primarily solves the fundamental problems associated with driving a densely populated backplane; as a result, it provides significant improvements in both speed and data integrity.[32] A detailed discussion on this subject can be found in Ref. 32. Here, the features of Futurebus transceiver will be briefly discussed in connection with noise immunity and EMI.

A solution which is now part of the IEEE P896 Futurebus is the low output capacitance of the transceiver. Using a Schottky diode in series with an open-collector driver output, the capacitance of the drive transistor is isolated by the small reverse-biased capacitance of the diode in nontransmitting state (Fig. 5.12). The Schottky diode capacitance is typically less than 2 pF and is relatively independent of the drive current. Allowing for a receiver input capacitance of another 2 pF, the total loading of the Futurebus transceiver can be kept under 5 pF. The P896 draft specification calls for a maximum plug-in capacitance of 10 pF.

The Futurebus transceiver has a precision receiver threshold centered between the low and high bus levels of 1 and 2 V, respectively, and provides a maximum percentage noise margin with respect to these bus voltage levels. A reduction of crosstalk also occurs because of the trapezoidal output waveform with a 6 ns transition time (See Fig. 5.12a). Besides reducing crosstalk, the controlled rise and fall times on the bus waveforms also reduce electromagnetic radiation from the bus. As representative Futurebus transceivers, the DS3896 is an *octal* device with common control signals, whereas the DS3897 is a *quad* device with independent driver input and receiver output pins.

Regarding the bus termination, the P896 draft requires that the bus be terminated at both ends, with a single resistor of 39 Ω connected to an active voltage source of 2 V, as shown in Fig. 5.12b. This arrangement has a considerably lower power dissipation than a *Thevenin-equivalent* two-resistor termination connected to ground and the 5 V rail.

RFI/EMI Reduction

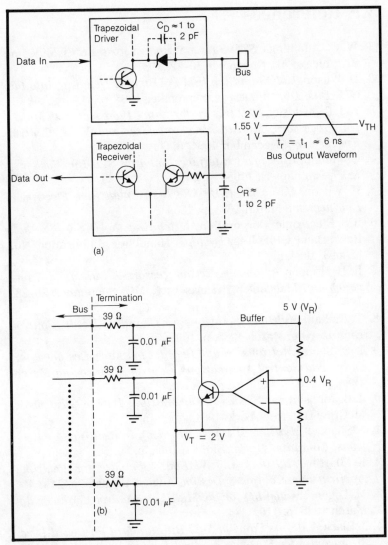

Figure 5.12—(a) Futurebus Trapezoidal Transceiver and (b) Futurebus Termination Circuit in a Backplane

5.8 References

1. W. C. Cummings, *STD Bus: A Standard for the 80's*, Midcom/81, Chicago, IL, November 10–12, 1981.
2. D. Wilson, *CMOS and the IBM PC Breathe New Life Into the STD Bus, Digital Design*, September 1984, p. 64.
3. A. J. Laduzinsky, *STD Bus Performance Increases With Multiple Processors, Multitasking Software, and VLSI, Control Engineering*, December 1984, p. 65.
4. C. Philipp, *Intelligent Interfaces Enhance STD Bus Systems, EDN*, September 5, 1985, p. 201.
5. H. Wilson, *Using the IEEE-488 Instrument Bus, Electronic Engineering*, March 1983, p. 155.
6. ICS Electronic Corporation, *Extending the IEEE-488 Bus*, Racal-Dana GPIB Test Systems Handbook, Publication No. 980593, 1984.
7. R. H. Hopper, *Adapting Microprocessors to Industrial Environments*, Maecon/83, Kansas City, MO, September 26–28, 1983.
8. T. Kinhan, *Multibus: Solutions for Changing Needs, Control Engineering*, March 1985, p. 108.
9. I. S. Shair, *Multibus Single-Board Computer Designed for Industrial Control Applications, Control Engineering*, March 1985, p. 120.
10. J. B. Johnson and S. Kassel, *The Multibus Design Guidebook*, McGraw-Hill, New York, 1984.
11. S. Shapiro, *Multibus II Makes Headway in Realtime Applications, Computer Design*, July 1, 1985, p. 35.
12. K. Martin, *Multibus and VMEbus Compare for Realtime Applications, Computer Design*, August 15, 1985, p. 24.
13. M. Logan, *Analog I/O for the Multibus, Control Engineering*, March 1985, p. 110.
14. J. Black, *VMEbus Family: A 32-Bit Standard for Board-Level Microcomputer Systems, Control Engineering*, August 1984, p. 112.
15. T. Balph and D. Artusi, *VME: A System Architecture for Industrial Control, Motorola System Design News*, Vol. 1, No. 1, March 1985, pp. 10–15.
16. J. Victor, *Top 32-Bit Buses Claim Multiprocessing Edge, Mini-Micro-Systems*, September 1985, p. 134.

References

17. M. J. McGowan, *From the S-100 to CAMAC: The Diversity of Digital Buses, Control Engineering*, April 1979, pp. 31–34.
18. J. Black and J. Kister, *VERSAbus: A Powerful Structure for Multiprocessing Applications*, Midcon/81, Chicago, IL, November 10–12, 1981.
19. Product Focus, *Bus Orientated Boards, Electronic Engineering*, July 1985, p. 70.
20. F. Tracy, *VLSI Solutions for Distributed Control: The BITBUS Interconnect and Distributed Control Modules*, Automach Australia '85, Melbourne, July 1985.
21. D. Head, *Testing MIL-STD-1553B, Communications Engineering International*, April 1983, p. 45.
22. Texas Instruments, *NuBus Specification*, Irvine, CA, 1983.
23. P. L. Borrill, *Futurebus: The Ultimate in Advanced Systems Buses*, Mini/Micro Northeast/85, New York, April 23–25, 1985.
24. R. Dalrymple, *Eurocard/DIN Single-Board Computers Signal Era of International Standards*, Mini/Micro Systems, August 1984, p. 171.
25. G. MacNicol, *Building Graphics Systems from the Board Level, Digital Design*, August 1984, p. 46.
26. R. L. Mack, *The PC/STD Connection*, Wescon/84, Electronic Show and Convention, Anaheim, CA, October 30—November 2, 1984.
27. Product Focus, *Bus Oriented Boards, Electronic Engineering*, July 1985, p. 70.
28. S. Bloom, *Serial Digital Bus Heads for Industrial Systems, Electronic Design*, September 13, 1980, p. 137.
29. C. J. Georgopoulos and C. Ross, *Digital Data Transmission System*, U.S.A. Patent No. 3,849,672, November 19, 1974.
30. C. J. Georgopoulos, *DC Noise Improvement in TTL Bus Organized Data Transmission Systems*, Proc. 1977 IEEE International Symposium on Circuits and Systems, Phoenix, AZ, June 21–23, 1977.
31. G. Connor, *Beat RFI and EMI With Differential Backplane Transceiver, Digital Design*, July 1984, p. 110.
32. R. V. Balakrishnan, *The Proposed IEEE 896 Future Bus—A Solution to the Bus Driving Problem, IEEE Micro*, August 1984, pp. 23–27.

General References

- J. Victor, *PC Bus Board Target STD Bus Markets*, Mini-Micro Systems, June 14, 1985, pp. 15–20.
- G. C. Roop, *Why Use Intelligent STD Bus Boards*, Control Engineering, December 1984, pp. 74–75.
- N. Laengrich, *Instrument Intelligence Determines 488 Bus Speed*, Electronic Design, October 14, 1982.
- N. Urbana, *Timing Considerations of GPIB Systems*, Wescon/80, Anaheim, CA, September 16–18, 1980.
- R. M. Williams, *LSI Chips Ease Standard 488 Bus Interfacing*, Computer Design, October 1979, pp. 123–131.
- A. Shuen and J. Beaston, *Multibus II Designs Exploit Advanced Concepts*, Computer Design, February 1985, p. 171.
- T. Doone, *Architecture Improves Multibus Performance*, Control Engineering, March 1985, p. 122.
- H. J. Hindin, *Bus Selection for 32-bit Limited to Two Choices*, Computer Design, September 15, 1985, p. 23.
- T. Naegele, *Newcomer Boards VME Bus*, Electronics Week, March 4, 1985, p. 61.
- P. Harold, *Powerful Local Buses Join VME Bus*, EDN, April 18, 1985, pp. 199–208.
- A. J. Laduzinsky, *AC Power Line Data Bus Delivers Error Free Digital Communications*, Control Engineering, October 1984, pp. 91–92.
- E. Strachar, *Hybrid Trio Takes Charge of the Bus-to-CPU Interface in Military Avionics Gear*, Electronic Design, April 4, 1985, p. 227.

Chapter 6
Data Acquisition Systems Interfaces

Data acquisition and distribution systems interface between the real world of physical parameters, which are analog, and the artificial world of digital computation and control. This chapter focuses on the basic building blocks that constitute a *data acquisition system (DAS)* and the problems associated with the various interfaces and interconnecting links when operating in an industrial environment.

6.1 Introduction

Most phenomena in nature are in analog form. Once translated into digital information, the data may be stored in either raw or processed form; it may be retained for some period of time or it may be transmitted over long or short distances. For collecting and converting analog information into digital form, a DAS is used. Input signals to a DAS are defined by their signal level source/load impedance, signal/noise frequency spectrum and environmental specifications (temperature, humidity, shock vibration, etc.).

Real world transducers produce signals which are highly subject to interference due to their small amplitude, extreme source impedance or unusual form. These signals must be amplified, impedance-matched, filtered and otherwise transmitted and transformed to the appropriate place and form for analog-to-digital (A/D) and then digital-to-analog (D/A) conversion. Remote inputs

DAS Interfaces

should be protected from overvoltage to prevent destruction of the equipment. Care should be taken to reduce RFI and EMI influence on input system signals as well as on relaying of output signals.

6.2 Basic Building Blocks of a Data Acquisition System

In this section, the basic building blocks which make up a typical data acquisition system are discussed along with their interface and interconnection requirements.

6.2.1 Data Acquisition System Architecture[1,2]

A typical data acquisition system (DAS) consists of six fundamental blocks (see Fig. 6.1):
1. An input *multiplexer* to select and transmit more than one signal over a single line for sampling,
2. A *programmable-gain amplifier (PGA)* to boost the level of the selected input as required to more fully utilize the analog-to-digital (A/D) converter's range and resolution,
3. A *sample and hold (S/H)* amplifier that functions like a digital buffer register; upon command, it "takes in" an analog signal and "holds" it until the A/D converter is ready for a new value,

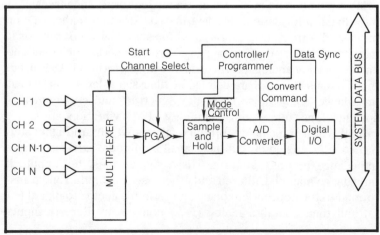

Figure 6.1—Block Diagram of a Typical Data Acquisition System (DAS)

4. An *analog-to-digital (A/D)* converter to take analog voltages or currents and transform them into digital logic levels,
5. A *timing system* needed to synchronize and coordinate the various functions of the DAS and
6. A *digital I/O port* to communicate between the *peripheral controller* and DAS.

6.2.2 Basic Interface Design Problems

Except for pre-amplification, all processing functions in a DAS can be performed digitally after analog-to-digital conversion. The low cost of digital ICs and the increase in chip complexity make digital processing easier and more cost effective compared to many analog functions. Typical examples include *microprocessors* with arithmetic logic and *read only memories (ROMs)*. The approach to interface design depends on the input signals and the type of microprocessor.

The interface block shown in the dotted lines in Fig. 6.2 indicates three types of problems when designing an interface.[3] These are data acquisition and distribution problems, including interconnections, converter(s) problems and processor problems.

The task of the designer here is to look at the limitations of each and make tradeoffs between cost, speed, accuracy and interconnection difficulties. In the following subsections, the various

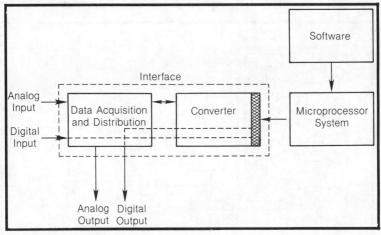

Figure 6.2—Interface Block Diagram within a Typical DAS [3]

DAS Interfaces

functional blocks of a DAS will be briefly discussed along with some problems pertaining to the first two categories of the above problems.

6.2.3 Input Multiplexer

A multiplexer switch connects selected analog input channels to the data-processing system. Typically, these switches are field-effect transistors (FETs) or equivalent devices in integrated form. These multiplexer chips also contain decoding and level-shifting drive circuits for operation, with standard TTL, ECL and CMOS logic circuits supplying the channel-address input.

A system can be configured to accept single-ended inputs and to digitize each channel, one at a time, by a means of an A/D converter. An alternative system configuration is that which can accept differential rather than single-ended inputs. In Fig. 6.3, for example, each input is multiplexed one at a time into high impedance, unity gain buffer amplifiers and then into a differential-to-single-ended amplifier.[4] Differential systems of this type are normally specified to accept signal (E_{SIG}) plus *common-mode* (CM) inputs of up to ± 10 V, i.e.,

$$CM + E_{SIG} = \pm 10 \text{ V} \quad (6.1)$$

This means that 5 V signals, subjected to ± 5 V CM or 1 V signals subjected to ± 9 V CM due to grounding differences, may be

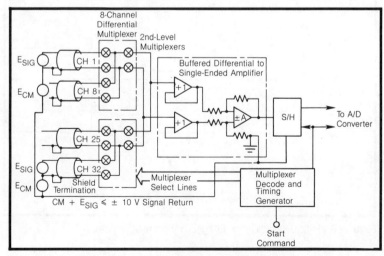

Figure 6.3—Differential Input Multiplexer (Adapted from Ref. 4)

considered typical differential plus CM ranges. As these voltages are multiplexed into the system, the unwanted CM signal is removed by buffered differential amplifiers while the differential signal is passed on to the A/D converter to be digitized. Input impedance and input current leakages are of prime concern, especially when considering a multiplexer for a measuring system.

6.2.4 Programmable Gain Amplifier

A 12-bit converter is theoretically capable of resolving 2^{12} steps within some full-scale voltage range; therefore, if a converter has a full-scale specification of 10 V, each bit of the digital output corresponds to $10/2^{12}$ = 2.44 mV. If the input signal to the DAS had a full-scale output voltage of 1.25 V for example, then the A/D converter would only be capable of resolving that input to within 2.44 mV, or one part in about 500. Such a conversion would therefore provide a precision of only about 9 bits. Using a gain amplifier to amplify the selected input by 8 (2^3) for example, will restore the full 12-bit resolution by *scaling* the input signal.

Using a *programmable gain amplifier (PGA)* with the selection of an input, a gain is also selected and results in a dynamic scaling of the input signal to more totally occupy the resolution range of the A/D converter.[2] Thus, a 12-bit A/D converter preceeded by a PGA, which can amplify an input signal by 64 (2^6) for example, will have an input-referred, full-scale range of 10/64 = 156 mV. Of course the system's resolution has not been extended beyond 12 bits, but its dynamic range has been increased to 12 + 6 = 18 bits.

6.2.5 Sample and Hold Amplifier

A sample and hold (S/H) amplifier is a form of an analog memory unit which ideally stores *(the hold)* an instantaneous signal *(the sample)* value at a desired point in time. A simplified diagram is shown in Fig. 6.4a. A sample command pulse momentarily closes the switch which allows the holding capacitor from leaking off.

DAS Interfaces

The time to charge the holding capacitor is called *acquisition time*. Normally, nine RC time constants are needed to charge the capacitor to 99.9 percent of the analog input.[5] The uncertainty of the actual point in time in which the switch is opened is called the *aperture time*. The voltage output change with time after the signal has been sampled is the *drift voltage*. Another constraint is *settling time*. This is the time necessary for the output buffer amplifier to settle at the new value. All of these constraints must be considered when selecting a sample and hold unit. The sum of the acquisition and settling times will determine the maximum sampling rate:

$$= \frac{\text{Sample Rate (max)}}{\text{Acquisition time + Settling time + Hold time}} \quad (6.2)$$

Errors are introduced by the drift rate and aperture time. Errors due to drift rate are rather straight forward, being expressed as voltage change per hold time interval. Errors due to aperture time depend on the signal rate of change at the sample point, as shown in Fig. 6.4b

With a finite aperture time, the voltage "held" can have changed as follows:

$$E = \frac{dv}{dt} t_a \quad (6.3)$$

where (dv/dt) = maximum signal rate of change
t_a = aperture time
E = voltage change of error

A/D Converters

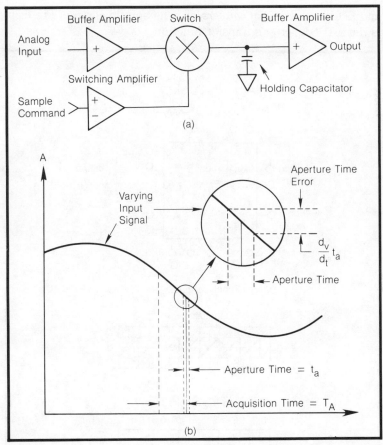

Figure 6.4—(a) Simplified Sample and Hold Circuit and (b) Aperture Time Error

6.2.6 Analog-To-Digital Converters[6-8]

The *analog-to-digital (A/D)* converter is the key component in the analog interface. An A/D converter is a device that accepts an analog input signal which may have any value falling between the minimum and maximum of the rated input range and which generates a digital output signal (a coded set of "1"/"0" levels). Figure 6.5a is the "black box" symbol for an A/D converter. Ideally the output is the precise digital representation of the analog output to within ±0.5 LSB, as shown by the top curve of Fig. 6.5b; LSB is the *least significant bit*, MSB is the *most significant bit*

DAS Interfaces

and EOC means *end of conversion*. The bottom curve of Fig. 6.5b is the A/D converter's transfer function.

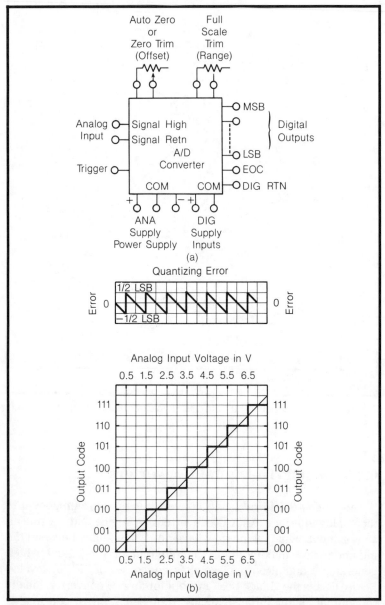

Figure 6.5—(a) "Black Box" Representation of an A/D Converter, and (b) Theoretical Transfer Function of an A/D Converter (First Three LSBs Only)

Most A/D converters are either *integrating* converters or *successive-approximation* converters. Integrating converters are inexpensive and accurate. The integrating conversion process has relatively low noise sensitivity but is much slower than other types. Successive-approximation converters provide faster conversion at moderate cost.

In the process of selecting an A/D converter, the following considerations are typical:
1. What is the analog input range, and to what resolution must the signal be measured?
2. What is the requirement for linearity error, relative accuracy, stability of calibration, etc.?
3. To what extent must the various sources of error be minimized as environmental temperature changes? Are missed codes tolerable under any conditions?
4. How much time is allowed for each complete conversion?
5. How stable is the system power supply? How much error due to power-supply variation is tolerable in the conversion system?
6. What is the character of the input signal? Is it noisy, sampled, filtered, rapidly varying, slowly varying? What kind of preprocessing is to be (or can be) done that will affect the choice and cost of the converter?

6.2.7 Digital-To-Analog Converters[3,4]

In a microprocessor-based system, as shown in the interface block diagram of Fig. 6.2, it will be necessary to convert the signals after processing back to analog form with a *digital-to-analog (D/A)* converter. D/A converters enable computers to communicate with the outside world. They are used in CRT display systems, voice synthesizers, automatic test systems, digitally controlled attenuators, process control actuators and other applications. They are also key components inside most A/D converters.

A D/A converter is a device that accepts digital signals (coded sets of "1" and "0" levels) at its input terminals and generates an analog current or voltage output that is, ideally, exactly representative of the input code, were each bit to be perfectly weighted. Figure 6.6a is the "black box" symbol for a D/A converter, and Fig. 6.6b is part of a typical transfer function, i.e., the transfer function for the three least significant bits of a binary input code. Most D/A

DAS Interfaces

converters operate on current-summing principles. In practice, D/A switches are in multiples of four, called quads. D/A converter specifications resemble those of A/D converter specifications.

There are four main criteria when selecting a D/A converter:
(1) resolution, accuracy, linearity and monotonicity characteristics;
(2) speed requirements and type of output (voltage or current);
(3) logic levels and type of digital code available for conversion and

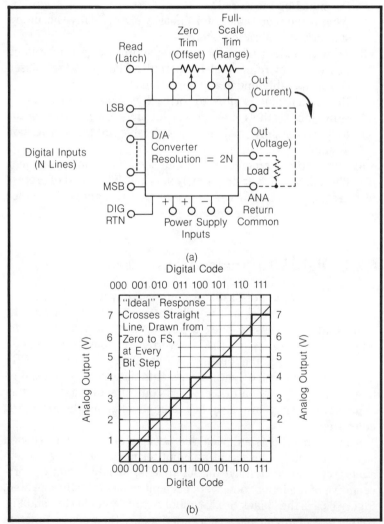

Figure 6.6—(a) "Black Box" Representation of a D/A Converter and (b) Theoretical Transfer Function of a D/A Converter (First Three LSBs Only)

(4) operating temperature range. Exactly what to look for depends on one's type of application.

6.2.8 Timing Considerations

Most modern DAS boards allow the *timing/control* functions to be performed by some appropriate *central processing unit (CPU)*, usually mounted on another card. Communication between the two devices will likely involve some standardized data bus, like the STD bus or the Multibus (see Chapter 5).

As demand for higher-speed operation increases, it becomes necessary to implement a significant amount of local intelligence on the DAS card itself. In particular, the following must be provided: registers for controlling analog functional modes (single-ended/differential, unipolar/bipolar), a memory device (RAM) capable of generating a continuous flow of gain/channel-select data, an address generator for the RAM, a pointer register for the RAM address generator to control sequence size, an interrupt generator to alert the system data bus that data is waiting and a bi-directional data bus interface to allow the writing of registers and RAM as well as verification of their contents.[2]

6.2.9 DAS-To-Microprocessor Input/Output Interface[9,10]

The interface requirements between the DAS and any microprocessor consist of three basic building blocks which make up the parallel interface: the *data bus buffers*, the *address decode logic* and the *handshake and control* circuitry. To communicate effectively with the DAS, the microprocessor always needs these three blocks, implemented by merely using a single LSI parallel interface element (provided by each microprocessor manufacturer) or by using several MSI discrete logic packages.

Interfacing A/D converters with microprocessors depends primarily on the A/D converter data rate. The key to simplifying the interface is to recognize the I/O category that is the most appropriate for the converter. Slow A/D converters are best interfaced to the microprocessor as *isolated I/O*. Self-contained peripheral interface circuits providing the required control and interrupt handshaking signals are available to implement the isolated I/O interfaces, primarily in applications where conversion times range from 1 μs to 10 ms. For applications with conversion times faster than 1 μs, *direct memory address (DMA)* must be used.

When designing the interface between a data acquisition system and a microprocessor, analog and digital *ground routing* in the DAS and noise emissions from the microprocessor-based system are often overlooked. Since the DAS depends on highly accurate voltage levels, maintaining that accuracy in an environment of large digital ground currents and noise is imperative. Care must be taken not only when laying out the printed circuit board, but also the analog section must be isolated as much as possible from its digital surrounding. This can be achieved by physical separation and considerable areas of ground plane, both analog and digital.

It is good practice for the digital and analog ground routing to be of two separate networks, originating at the system power supply as one common ground. The analog ground path back to the system power supply must contain little or no varying currents. At the same time, it should be as mechanically short and thick as possible to keep a constant analog ground reference throughout the entire system. The separation between the two ground networks should not stop at the system bus, but should continue through every card in the system. Clamping diodes, installed between the two grounds on cards containing the analog and digital circuitry, prevent damage to circuitry in the event of accidental separation between the two grounds. (See also subsection 6.4.3.)

6.3 Data Acquisition and Data Interface Considerations

Accurately transmitting the output signal from a transducer across some distance to a central location for display, digitization or computation and providing error-free input signals to remote actuators is a persistent problem for process control and similar industrial environments.

6.3.1 A Microcomputer-Operated Process Control System

Systems known as computer analog I/O boards have become quite popular. Since the computer interface to a data acquisition system is always a major problem, a logical solution is to design the data acquisition system into a board that plugs directly into the microcomputer or its housing. The concept of placing the A/D and D/A converter components "inside" the computer was not practical before the development of the modern high-performance, 12-bit hybrid and monolithic A/D and D/A converters. These new miniature devices make it possible to include A/D converters, analog multiplexers, sample-hold circuitry and D/A converters, together with required interfacing logic, on a compact circuit card.[11] A microcomputer-operated process control system is shown in Fig. 6.7.

By applying the above concept, two major application problems that have hampered low-cost data acquisition designs for years are solved. First, it relieves the user of risky, tedious interface circuit design by offering a proven production design. Second, it eliminates the expensive digital cabling required for connecting to external A/D and D/A components. Only analog signal cabling is required. However, the majority of existing and future systems do not fall into the above category. Therefore, careful consideration should be given to not only to the analog cable, but also to complete links from sensors/transducers to the system as well as from the output of the system to various actuators and other devices.

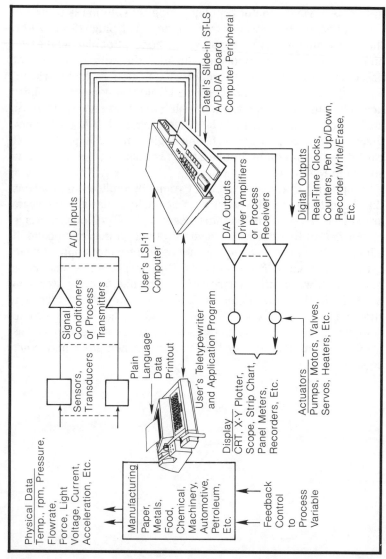

Figure 6.7—A Microcomputer-Operated Process Control System (Reproduced from *Digital Design* by Permission of Morgan-Grampian Publishing Co.)

6.3.2 Source Characteristics

In the design or selection of components for the data acquisition part of a large system, consideration should begin with the *sensors* and *transducers*, i.e., the *signal sources*. The basic *transducer* concept is shown in Fig. 6.8 and can be divided into[12] the sensor and sensor block, analog signal conditioning circuits and buffer and feedback control elements.

The sensor is an element that dips into a physical environment and collects data within a limited range or area which, in turn, is used as control information. Thus, a special physical-to-electrical or electronic transfer element called the sensor is used to change data of physical phenomena into electrical signals. The combination of sensor and high-gain amplifiers is often called the *sensor block*. The sensor signal is conditional, and is adapted in successive signal conditioning elements until it can be fed into the data link. The chain of sensor, high gain amplifier and signal conditioning elements is called *transducer*.

Some of the most important sources, or transducer characteristics, that should be taken into consideration are:

1. Signal level available and source impedance,
2. Transfer accuracy of source itself,
3. Desired dynamic amplitude range,
4. Anticipated change in source transfer characteristic as a function of environment (temperature, humidity, etc.),
5. Required stability of source power supply,

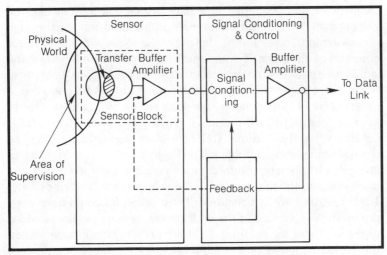

Figure 6.8—Basic Transducer Concept (Adapted from Ref. 12)

6. Required frequency of source monitoring to provide meaningful data and a minimum of unnecessary data,
7. Maximum acceptable distance between source and acquisition system,
8. Location of source ground (common), if different from system ground,
9. Required amplification at the source (if any),
10. Source output: single-ended or differential and
11. Probability of power line and noise pickup.

The data link interconnects the transducers with the data processor. In industrial environments, special modulation techniques may be used to minimize the influence of high noise levels. Two representative interconnection schemes between a source (sensor and transducer) to a measuring or data collecting system will be discussed in the next two subsections.

6.3.3 Data Acquisition From Thermocouple Inputs[13]

Thermocouples are often used as temperature sensors for process control systems. Thermocouples are characterized by temperature coefficients of 10 to 70 μV/°C and operating ranges of minus hundreds to plus thousands of degrees centigrade. The wires running from the thermocouple to the measuring device often pick up large *common-mode noise* signals of 60 Hz or higher frequencies. For example, in a data acquisition system like the Burr-Brown MP22, which has eight differential inputs, the high common-mode rejection of the *instrument amplifier* will reject common-mode noise. To minimize *differential mode noise*, the signal wire should be twisted and, if possible, shielded. In applications where these wiring practices cannot always be observed, a *differential RC filter* may be used, as shown in Fig. 6.9a. The 10 kΩ resistors and 10 μF capacitor provide low pass filtering (f_c = 0.8 Hz), while the optional 1 MΩ resistors supply bias current to the instrumentation amplifier. The *remote sensor* should be earth grounded to prevent common-mode voltages from exceeding the ±5 V range of the multiplexer. If the sensor is earth grounded, the 1 MΩ resistors are not required. These resistors could have been put on the output side of the multiplexer, eliminating the need for repeating them for each input; however, this would have loaded the 10 kΩ resistors of the filter causing a possible one percent error for static conditions.

Thermocouples

To complete a thermocouple system it is necessary to terminate all thermocouple wire pairs at an *isothermal box* or connector strip of some type. An ordinary *barrier strip* may be monitored to allow the observed thermocouple *electromotive force (EMF)* to be cold junction compensation. A very suitable circuit for this purpose is shown in Fig. 6.9b. Its output sensitivity is approximately 2 mV/°C, and the holding relationships are:

$$E = \frac{Rd}{Rc + Rd}\left(1 + \frac{Ra}{Rb}\right)\frac{kT}{q}\ln 10 \qquad (6.4a)$$

$$\left(\frac{dE}{dT}\right) = -\frac{Rd}{Rc + Rd}\left(1 + \frac{Ra}{Rb}\right)\frac{K}{q}\ln 10 \qquad (6.4b)$$

where $T = °K$, $k/q = 8.67 \times 10^{-5}$

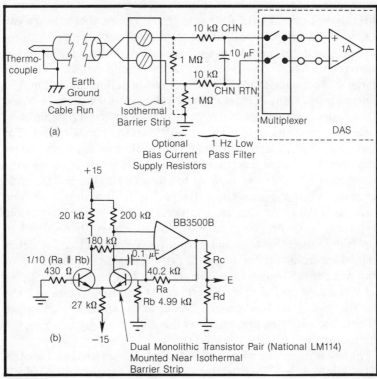

Figure 6.9—(a) Thermocouple to DAS Connections and (b) Monitoring Circuit

6.3.4 Voltage-To-Frequency and Frequency-To-Voltage Converter Links

Voltage-to-frequency (V/F) converters and frequency-to-voltage (F/V) converters can be used to solve tough data acquisition problems. These versatile subsystems can transmit data from analog sensors over simple two-wire links and minimize the effects of electrical noise and transients on the transmitted signal. By connecting a V/F converter with a counting circuit, computer-compatible digital signals can be generated. By doing the reverse and feeding pulses to an F/V converter, analog signals can be generated, which in turn can control a process or instrument.

Various modular V/F converters are available which handle signals from millivolts to well above 100 V. Both V/F and F/V converters are available, with various frequency ranges from 1 kHz to well over 5 MHz, full scale. Furthermore, these ranges can be modified with external circuitry to match specific applications.

Compared to successive-approximation or tracking-type A/D converters which can give completely wrong output codes for large spike inputs, the V/F converter has high noise rejection. The output pulse train is counted over relatively long periods. Noise that appears as a repetitive waveform is averaged out completely over the measuring interval, except for fractional cycles. For example, if the interfering signal is a 60 Hz sine-wave, the worst case contribution to the output which occurs when the measuring interval is 1/2-cycle too short or too long is 1/120 of the 1/2-cycle average when the measuring interval is in the vicinity of one second. If the count time is an integral multiple of the line period, rejection is theoretically infinite, and is better than 80 dB in practice. Since the counting period for V/F converters is determined externally, it can be set to provide maximum rejection for the most troublesome noise frequencies. For example, multiples of a counting period of 0.1 Hz will reject line frequencies of 50 or 60 Hz. And if the fundamental has an integral number of cycles during the counting period and is therefore strongly rejected, the harmonics will also be rejected.[14]

Figure 6.10 shows a V/F converter link using the AD537 device. The AD537 is a V/F converter in integrated circuit form and is particularly well-suited to remote data acquisition applications. It includes a high impedance buffered input stage, low power consumption and versatile output stage. The low power requirements of the AD537 permit use with a simple two-wire, twisted-pair link which carries power to, and frequency data from, the V/F converter. A simple current monitoring circuit like the one shown in Fig. 6.10 can be used to recover the frequency information. This circuit has the added feature of a stable 1 V reference available at the remote location. This voltage can be used to provide excitation for a resistive transducer.[15]

On the other hand, periodic events abound in nature and information concerning their periodicity can be extracted using an F/V converter. Figure 6.11 shows the use of an F/V converter

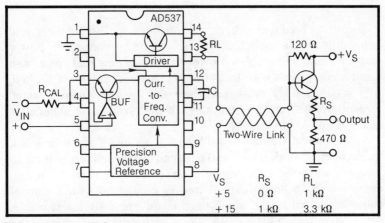

Figure 6.10—V/F Converter Link Using the AD537 Device

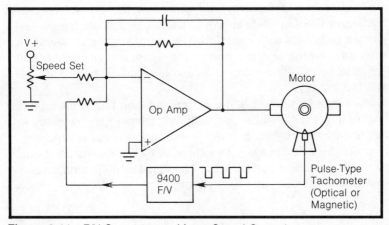

Figure 6.11—F/V Converter as Motor Speed Control

DAS Interfaces

operating as a stand-alone device for rpm measurement control. It converts the signal from a pulse train output rpm transducer (tachometer) to a voltage that either appears on a meter or strip chart, or controls the shaft speed.

6.3.5 Single-Line Power and Data Transmission Link

The use of common transmission lines for both power and communication signals has been an attractive concept to power and telephone companies. However, the systems that have been developed are two-wire transmission systems which are limited; they are extremely noisy, and the noise level fluctuates greatly. Noise derives from the generation of the power, line impedance changes and load variations.

A solution to the above problems is to use a single-line link through which both power and data can be transmitted, using a 15 V supply and pulse width modulation (PWM) technique.[16] The link consists of two basic units, A and B, connected with a coaxial cable as shown in the block diagram of Fig. 6.12. Local unit A sends quantized power levels along with control signals to the remote unit B. At unit B the quantized power is applied to a voltage regulator after being rectified and filtered, the output of which forms a "temporary" dc power in the remote section. Signals, or data, sensed via proper transducers at the remote end (B) are converted into digital form and transmitted to section A where measuring or other processing equipment are available. It can also be readily converted to a voice channel by proper A/D and D/A conversion.

The single-line power and data transmission link is useful to a host of industrial applications, including temperature, pressure or position measurements; security fire-alarms; process or supervisory control; etc. It can easily be expanded by developing additional modules to perform such functions as signal amplification, analog-to-digital conversion and multiplexing.

Alternative Techniques

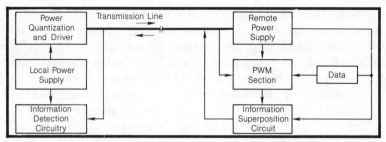

Figure 6.12—Block Diagram of Single-Line and Data Transmission Link (Adapted from Ref. 16)

6.4 Protection of Data Acquisition Systems from Interference

When designing an industrial process-control system, the designer is charged with delivering very small signals reliably through hostile environments over long distances and, as always, will be asked to do it as simply and inexpensively as possible. Equipment located at remote sites is especially vulnerable to electrical and other hazards. Remote inputs should be protected from overvoltage to prevent destruction of the equipment by accidental connection to the primary line. Care should be taken to reduce RFI and EMI influence on system signals.

6.4.1 Alternative Techniques for Voltage Isolation and Analog Data Transmission from Remote Transducers

One area of particular interest is the acquisition of data at sensors distant from the place where the data are required. A similar problem is the control of actuators at a distance from the source of the control signal. Solutions to these problems have existed from the time of the first large installation. A crucial link in the chain of components that form an acceptable solution is the signal transmitter/receiver scheme. It influences the number of transmission lines, the transmission type and, ultimately, the system's overall noise immunity.

Using a paired V/F and F/V scheme, for example, offers a wide range of transmission choices[17], as shown in Fig. 6.13. As was

DAS Interfaces

mentioned previously, the V/F and F/V converters can be used to solve tough data acquisition problems. The combined A/D converter/data links of Fig. 6.13 include: the use of fiber optics and optical isolators,[18,19] infrared light (IR),[20,22] ultrasonic waves,[21] magnetic tape or copper wire as the transmission medium.

The use of fiber optics and optical isolators will be discussed in Chapter 8. The next three links do not include wires; consequently, they will not be discussed further. The copper-wire link and its associated problems will be discussed in some detail. Figure 6.14 graphically shows, in a simplified way, the problems of interconnections in the presence of a typically hostile electromagnetic environment, together with heating effects and aging effects.[23] After realizing the problems of interconnections, one must consider electromagnetic and electrostatic pickup, voltage isolation and power-supply coupling, hazardous areas and heating effects. It should be very clear that the switching action of one building block may induce spurious signals into another building block unless those building blocks are properly designed and the appropriate precautions are taken.

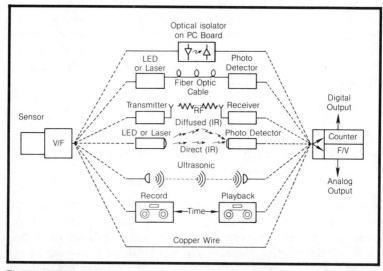

Figure 6.13—Alternative Techniques for Voltage Isolation and Analog Data Transmission from Remote Transducers, Using V/F and F/V Converters

Input Conditioning

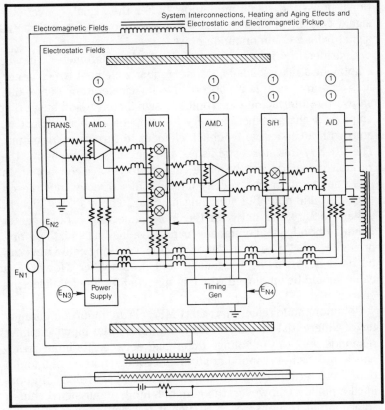

Figure 6.14—Effects of Hostile Environments on Data Acquisition Systems (Reprinted from *EDN*, December 15, 1972, © Cahners Publishing Co.

6.4.2 Input Amplifiers and Multiplexer Section Conditioning

Real-world transducers produce signals which are highly subject to interference because of their small amplitude, extreme source impedance or unusual form. These signals must be amplified, impedance-matched, filtered and otherwise transmitted and transformed from the condition in which they are produced to the appropriate place and form for the A/D converter. An amplifier and a filter are critical components in the initial signal processing. The amplifier must perform one or more of the following func-

DAS Interfaces

tions: boost the signal amplitude, buffer the signal, convert the signal current or charge into a voltage or extract a differential signal from a common-mode noise.

Consider, for example, the piezoelectric accelerometer, a device employing a mass-loaded ceramic or quartz element which generates an output charge proportional to its acceleration. Figure 6.15 shows the interfacing electronics: a signal-conversion stage that employs a charge-sensitive, very low impedance bias-current op amp with a capacitive feedback element to furnish a V/g output. The 10^{11} Ω resistor establishes the circuit's dc stability; the variable capacitor sets the stage's charge-to-voltage conversion. In this configuration, the accelerometer looks directly into the op amp's ground-potential summing junction. The resulting lack of voltage differential between the interconnecting cable's center conductor and shield eliminates capacitive loading and, therefore, prevents signal-time degradation. In this way, long cable runs can be used between the transducer and its amplifier. However, a cable should be employed which is specified for low triboelectric-charge effects.[24]

Common-mode rejection ratio (CMRR) is an important parameter of differential amplifiers. An ideal differential input amplifier responds only to the voltage difference between its input terminals and does not respond at all to any voltage that is common to both input terminals, i.e., to common-mode (CM) voltage. In non-ideal amplifiers, however, the common-mode input signal causes some output response even though it is small compared to the response to a differential input signal. The ratio of differential and

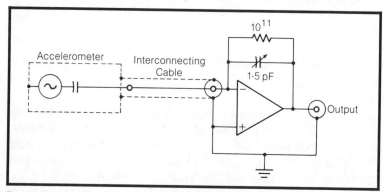

Figure 6.15—Piezoelectric Accelerometer Interface

common-mode responses is defined as a common-mode rejection ratio (in dB) as follows:

$$\text{CMRR} = 20 \log A_D/\text{Acm} \qquad (6.5)$$

where A_D is differential voltage gain and Acm is common-mode voltage gain. CMRR is a function of frequency and, therefore, also a function of the impedance balance between the two amplifier input terminals.

Figure 6.16a shows a balanced common-mode rejection equivalent circuit at the input of a differential amplifier driving an A/D converter. For better results, R_{S1} should equal R_{S2} and C_{S1} should equal C_{S2} in order to minimize the effect of common-mode voltage (E_{cm}). However, since most differential multiplexed A/D systems are not operated under ideal conditions, it is often better to ascertain the common-mode rejection under some given source imbalance at specific frequencies. Considering, for example, Fig. 6.16b, source imbalance may exist as shown in the equivalent circuit of Fig. 6.16c due to various resistance and capacitance components which contribute to common-mode degradation. If $R_{S_2} = R_{S_1}$ and $R_{IMB} = 1$ kΩ, the effective common-mode impedances, X_{CM_1} and X_{CM_2}, will be much greater than the source impedances, R_{S_1} and R_{S_2}. Thus, R_{IMB} and X_{CM_1} form a voltage divider causing a voltage to appear as a differential input to the amplifier; even though it is caused by E_{CM}, the amplifier cannot distinguish it from a true differential signal.

A significant reduction of the effects of node capacitance, and hence of the undesirable differential input signal, can be achieved by using submultiplexing (Fig. 6.3) and by observing the shield grounding rules. By properly terminating the shield at the source of the common-mode (Fig. 6.16b), signal-to-shield capacitance is effectively bootstrapped, thereby not affecting common-mode ratio. In addition, submultiplexing (a second level of multiplexing) reduces settling time, crosstalk and leakage current in a multiplexed A/D converter by minimizing the number of switches (FETs or equivalent devices) tied to the amplifier input.[4]

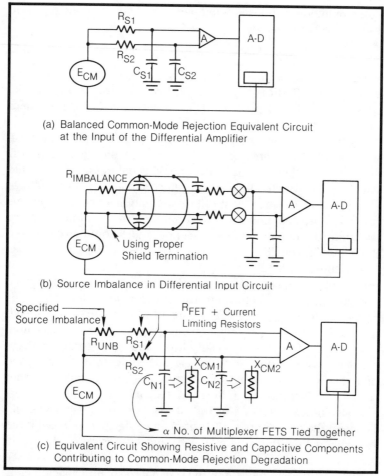

Figure 6.16—Input Amplifier Common-Mode Problems[4]

6.4.3 Problems With A/D and D/A Subsystems[25-28]

Modern fast converters are excellent building blocks for high-speed signal-processing or test and measurement systems. The application, however, can present some difficult problems, since poor grounding practices can lead to intolerable data-converter inaccuracies, improper bypassing can give rise to high D/A converter output noise, noisy D/A converter digital inputs can result in noisy outputs, changing ranges can seriously compromise a D/A converter's setting time, insufficient A/D converter driving-

A/D and D/A Problems

source capability can cause missing codes and real A/D and D/A converters exhibit a number of departures from the ideal transfer functions. These departures include offset, gain and linearity errors.

For example, in successive-approximation converters, the input current is compared to a trial current (Fig. 6.17). The comparison point is diode-clamped, but it may swing plus-and-minus several hundred millivolts, which gives rise to a modulation of the input current. Additionally, fast, high-resolution systems may suffer from amplifier output-transient errors. Solutions to these problems include: (1) the use of an A/D converter that has its own onboard buffer, (2) the use of a sample-hold, with low-output impedance, that can serve as a buffer, (3) the use of a wideband op amp which does not include output current-limiting resistors and (4) the construction of an inside-the-loop buffer that can stiffen the output of a slow, accurate amplifier.

Bypassing and decoupling of conversion components needs special attention. In *virtual ground* systems such as an op amp driven by a current-output D/A converter, the D/A output current does not actually return to ground. It returns to one of the power supplies, by way of the op amp's output stage (Fig. 6.18). To reduce the impedance in the high-frequency current path, the bypass capacitor should be connected so as to return the currents from one (or both) power terminals to ground at the D/A

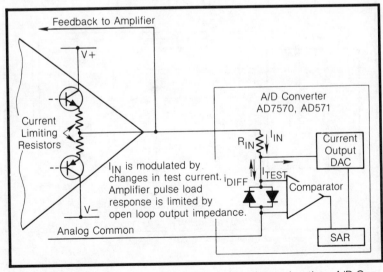

Figure 6.17—Relationship between Successive-Approximations A/D Converter and Op Amp That Is the Source of the Input Signal [26, 27]

DAS Interfaces

converter. If the converter output is active, it may require bypassing of its own supplies for the same reason.

An effective decoupling scheme is shown in Fig. 6.19. In this configuration, the decoupling capacitor connects the shortest path between the load return and the load-voltage control element. Here, an op amp swinging a resistive load-circuit negative drives the load from an internal pnp transistor connected to the negative supply. Decoupling the negative supply pin of the op amp to the low side of the load provides the most direct return path for high-frequency currents and bypasses them around ground and power buses.

In large systems, and in systems which deal with both high-level and low-level signals, ground or common bus management becomes an important aspect of design. As Fig. 6.20 shows, an analog subsystem can be locally interconnected with a single-wire connection to the digital common. This signal connection carries only the digital currents required for the converters's digital interface. Moreover, analog signals are not forced to share a conductor, even with those currents. The analog subsystem should be powered by a supply with a local common return which may be connected to the digital common but does not share any current-carrying conductors.

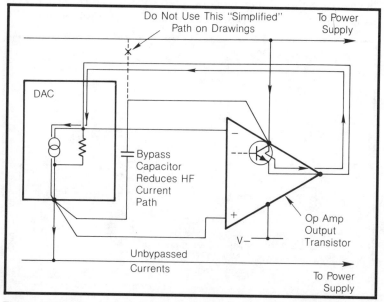

Figure 6.18—Bypassing poser supplies for virtual-ground applications. Arrows show unbypassed current flow.[26, 27]

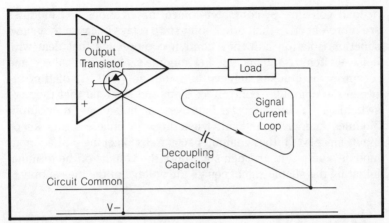

Figure 6.19—Decoupling Negative Supply Optimized for "Grounded" Load [26, 27]

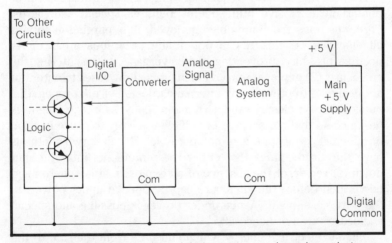

Figure 6.20—This connection minimizes common impedance between analog and digital (including converter digital currents).[26, 27]

6.4.4 Data Relaying Section

In a microcomputer-operated process control system (Fig. 6.7), the various relays that accept control signals must provide substantial electrical isolation between the loads they switch and the computer that supplies their control signals. Also the relays must be flexible enough to handle the computer's wide range of dc- or ac- switching voltages and to switch a wide range of ac- or dc-load voltages. In computer-control applications, I/O modules are more effective than other solid-state relays because they have superior isolation and they contain components that deal with noise switching signals. An I/O module's phototransmitter and receptor components achieve as much as 4,000 V isolation between the computer controlling them and the load that they are switching. A Schmitt trigger improves the noise margin in output modules and prevents fault triggering. An enable signal keeps input signals to the computer clean by keeping an ac-input module's output in an open state until the voltage of the terminal initiating the enable signal equals the voltage of the bias supply.[29]

6.4.5 Hazardous Areas and Temperature Effects

The use of remote acquisition equipment in explosive or potentially explosive atmospheres requires special safeguards. Three methods are commonly employed:[30] (1) purging maintains all electronic enclosures in potentially hazardous areas under positive pressure, preventing explosive gases from entering the enclosure, (2) explosion-proof enclosures allow gases to enter but are designed to prevent any internal explosions from propagating outside the enclosure and (3) intrinsically safe equipment is designed so that a discharge of sufficient energy to cause ignition of specified atmosphere is not possible. The first two methods have the disadvantage that either the remote site must be shut down for repair, or the absence of an explosive atmosphere must be verified before the enclosure can be opened under power. In the third case, maintenance under power is possible and special enclosures are not necessary.

Usually, electrical energy which can pass from the safe to the hazardous area is limited by an interface mounted in the safe area, the most common being the zener barrier. This safety technique is inherently low power and is really only useful for the protection of instrumentation and telemetry systems consuming less than one watt.[31]

A/D and D/A converters drift with temperature, as will all electronic equipment. Since wide temperature variations can be expected at remote sites, remote equipment must be designed for lower drift than equipment designed for control rooms. Often, offset drift can be removed by auto-zeroing; that is, measuring offset drift on an identical channel with its input short circuited and subtracting this offset from the other channels. Accuracy drift will normally be specified over some temperature range, for example, −25°C to +75°C. Operating temperature ranges depend on the particular application, as shown in Fig. 6.21.[32] Sometimes it may be important that equipment remain operational outside this range even with degraded accuracy. This is often not specified. As the electronics at most remote sites become more complex, heating or cooling of the enclosure becomes more economic than supplying wide temperature range components.[30]

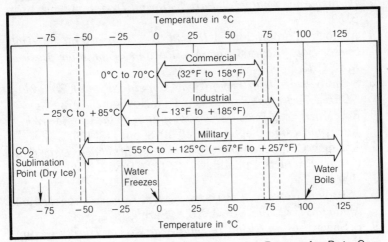

Figure 6.21—Standard Operating Temperature Ranges for Data Converters (Reproduced from *Digital Design* by Permission of Morgan-Grampian Publishing Co.)

6.5 References

1. L. Solomon and E. Ross, *Educating 'Dumb' Data Acquisition Subsystems*, Digital Design, November 1976, p. 29.
2. J. E. Leureault, Jr., *High-Speed Data Acquisition Design System*, Electronic Products Magazine, March 4, 1983, pp. 89–92.
3. N. Jagannatan, *Data Acquisition/Microprocessor Interface Performance Analysis—Part 1*, Digital Design, August 1979, p. 32.
4. Analogic Corporation, *A-D and D-A Converters for High Speed Data Acquisition Applications*, Computer Design, April 1973, pp. 57–64.
5. S. Muth, *Video Digitizing (Some Considerations)*, 73 Record NEREM Technical Papers (Part 1), November 1973.
6. R. L. Morrison, *Linking Microcomputers to the Outside World*, Machine Design, September 21, 1978, pp. 118–125.
7. Analogic Corp., *Designers Guide and Handbook: Data Conversion Products*, Bulletin No. BR 1021, Rev. 2, Wakefield, MA, 1982.
8. C. Brown, *Notes on Converter System Component Selection*, Analogue Dialogue, Vol. 6, No. 1, Spring 1972, p. 8.
9. E. Sliger, *Design Guide To Data Acquisition*, Digital Design, January 1981, p. 71.
10. D. C. Pinkowitz, *Achieve High Speed Data Acquisition with Fast ADC and DMA*, Digital Design, August 1983, pp. 103–107.
11. E. L. Zuch, *Principles of Data Acquisition and Conversion—Part 5*, Digital Design, March 1980, p. 68.
12. R. C. Richards and F. F. Stucki, *Interface and Data Format Converters in Automotive Data Acquisition and Control Systems*, Electronic Displays and Information Systems, International Congress and Exposition, Detroit, Michigan, February 23–27, 1981.
13. Burr-Brown, *Microprocessor—Interfaced 12-Bit Data Acquisition System*, Application Note PDS-387A, October 1978.
14. F. Pouliot, *Have You Considered V/F Converters?—They Offer High Resolution at Low Cost. Use Them for Digitizing, Isolating, Integrating, and Much More*, Analog Dialog, Vol. 9, No. 3, 1975, pp. 6–9.
15. D. Grant, *Novel Techniques for Precision Remote Data Acquisition*, Electro/81, New York, April 7–9, 1981.

16. C. J. Georgopoulos and G. A. Iordanidis, *Single-Line Power and Data Transmission Link*, 1985 Industrial and Commercial Power Systems Technical Conference, Denver, CO, May 13–16, 1985.
17. F. Goodenough, *Modular V/F's and F/V's: Simple Solutions to Everyday Conversion Problems*, Teledyne Philbrick Applications Bulletin AN-32, May 1977.
18. C. J. Georgopoulos and C. S. Koukourlis, *Fiber-Optic Link Design Considerations for Applications in Noisy Industrial Environment*, IEEE Transactions on Industrial Electronics, Vol. IE-31, No. 3, August 1984, pp. 209–215.
19. C. J. Georgopoulos, *Fiber Optic Sensors and Data Collection Highway Technique Applied to High Energy Systems*, ESS '83: Energy and Environmental Systems, Athens, Greece, September 1, 1983.
20. C. J. Georgopoulos, *The Use of Fiber Optics and Infrared Devices in Industrial Robots*, MECO '83: Measurement and Control, Athens, Greece, 1983.
21. C. J. Georgopoulos, *Alternative Communications Techniques in the Office of the Future*, HETELCON, Athens, Greece, August 1983.
22. C. J. Georgopoulos and V. C. Georgopoulos, *Hybrid IR/RF Transmission System With Common Modulating Circuit*, Electronics Letters 23, Vol. 20, No. 23, November 8, 1984, pp. 971–973.
23. B. M. Gordon and K. E. Jackson, *Watch Those Interactions in Your A/D and D/A Conversion Systems*, EDN, December 15, 1972, pp. 32–36.
24. J. Williams, *Exotic-Transducer Interfacing Call For Proven Techniques*, EDN, February 17, 1982, p. 195.
25. E. L. Zuch, *Interpretation of Data Converter Accuracy Specifications*, Computer Design, September 1978, pp. 113–121.
26. A. P. Brokway, *Analog Signal-Handling for High Speed and Accuracy*, Analog Dialogue, Vol. 11, No. 2, 1977, pp. 10–16.
27. D. J. Travers, *Precision Analog-to-Digital Converters Interface Techniques*, Wescon/80, Anaheim, CA, September 16–18, 1980.
28. B. Travis, *Follow Driving, Grounding Rules to Optimize ADC, DAC Operation*, EDN, March 17, 1983, p. 187.
29. K. Marrin, *I/O Boards and Software Give Cs Remote Control Over Solid- State Relays*, EDN, May 30, 1985, p. 79.
30. W. Archibald and Wiatrowski, *Remote Multiplexing*, Burr-Brown Research Corp., Application Note An-80, January 1976.

31. C. Burkitt, *Simplifying Apparatus for Hazardous Areas*, Control and Instrumentation, March 1985, p. 52.
32. E. L. Zuch, *Principles of Data Acquisition and Conversion—Part 4, Digital Design*, October 1979, p. 88.

General References

- F. J. Oliver, *Practical Instrumentation Transducers*, Hayden Book Co., New York, 1971.
- R. K. Hester, et al., *A Monolithic Data Acquisition Channel*, IEEE Journal of Solid-State Circuits, SC-18, No. 1, February 1983, pp. 57–65.
- J. Williams, *Fast Comparator IC Speeds Converters and S/H Amplifiers*, EDN, June 20, 1985, pp. 115—128.
- D. Mercer and D. Grant, *8-Bit A-D Converter Mates Transducers With Ps, Electronic Design*, January 12, 1984, pp. 361–367.
- P. Bradshaw, *A Few Chips Put a Transducer On a Two-Wire Serial Link, Control Engineering*, February 1980, pp. 57–60.
- R. Judd and T. Cataldo, *I/O Boards Bring Thousands of Data Points Under PC Control, Electronic Design*, September 5, 1985, p. 177.
- G. Clarke, *Protecting Plants Against Lightning Strikes, Control and Instrumentation*, May 1985, p. 59.
- D. Grant, *Attaining Microprocessor Interface Compatibility With DAC and ADC Devices, Computer Design*, December 1980, p. 159.
- Y-H. Sutu and J. J. Whalen, *The Sensitivity of Demodulation RFI Predictions in Op Amp Circuits to Variations in Model Parameters*, 1985 IEEE International Symposium on Electromagnetic Compatibility, Wakefield, MA, August 20–22, 1985.
- S. Connors, *Protect Data-Acquisition Systems With the Right Input Isolation, Electronics*, April 24, 1980, pp. 134–141.
- L. H. Sherman, *Sensors and Conditioning Circuits Simplify Humidity Measurement*, EDN, May 16, 1985, p. 179.

Chapter 7
Measurement and Automation Interfaces

Many electronic and electrical devices—from computers to wall plug-in shavers—generate electrical noise. This noise is either conducted along power cables to other electrical devices or radiated to the world outside, thereby producing interference that affects still other electrical instruments. Effective noise detection and measurement involves careful selection of instrumentation, proper design of test fixture interfaces and knowledge of the EMI test limits for compliance to standards.

7.1 Introduction

In a measuring process, weak signals at high impedance levels are enhanced by instrumentation amplifiers, where differential inputs make most noise pickup common-mode and self-canceling. Isolation amplifiers especially help designers solve problems in instrumentation, industrial monitoring, process control and patient monitoring.

Interface fixtures take several forms, but they may be generally defined as an assembly of wires and connectors that connect a unit under test (UUT) to the automatic test equipment (ATE) or tester. The acceptance of general purpose interface bus (GPIB) has accelerated the implementation of automated testing by significantly reducing the engineering costs of putting systems together. Testing computing devices to demonstrate compliance with FCC rules requires a knowledge of appropriate measurement techniques and test instrumentation for conducted and radiated emissions.

7.2 Instrumentation Amplifiers and Input Noise Problems

Industrial controls use transducers, such as thermometers and strain gages, which produce signals so weak that they can be masked by electrical noise. Instrumentation amplifiers can extract these microsignals more effectively than conventional operational amplifiers (op amps), but potential error sources may exist in the amplifier connections unless careful measures are taken.

7.2.1 Definitions and General Characteristics

In a measuring process, a physical parameter is first obtained by a transducer and then is amplified and filtered before it is processed further. An amplifier and filter are critical components in this initial signal processing. The amplifier must perform one or more of the following functions: boost the signal amplitude, buffer the signal, convert a signal current into a voltage or extract a differential signal from common-mode noise.

The most popular type of amplifier is an *op amp* which is a general-purpose gain block with differential inputs. The op amp may be connected in many different closed loop configurations: current-to-voltage conversion, inverting and noninverting voltage gain and unit gain buffer. In general, an operational amplifier is a good choice where a single-ended signal is to be amplified, buffered or converted from current to voltage.

An *instrumentation amplifier*, one of the key elements in a data acquisition or measuring system, is an op amp circuit used to measure small differential voltages riding on a common-mode voltage that is frequently larger than the differential voltage. Other common terms for this type of amplifier are *transducer amplifier*, *difference amplifier* and *bridge amplifier*.[1] Instrumentation am-

Definitions

plifiers have the following important characteristics:[2] (1) high-impedance differential inputs, (2) low-input offset voltage drift, (3) low-input bias currents, (4) gain easily set by means of one or two external resistors and (5) high common-mode rejection ratio (CMRR). Two representative instrumentation amplifier circuits are shown in Fig. 7.1. The first circuit is a simple single amplifier with voltage gain,

$$A_v = \frac{R_f}{R_1} \qquad (7.1)$$

The second circuit is formed from the first circuit by adding two follower amplifiers, which increase input resistance, and adjustable gain. Its voltage gain is:

$$A_v = 2\left(1 + \frac{R_f}{R_2}\right)\frac{R_f}{R_1} \qquad (7.2)$$

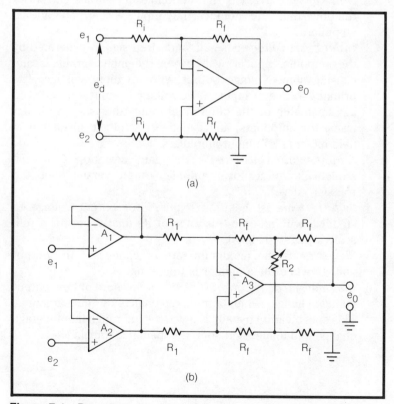

Figure 7.1—Representative Instrumentation Amplifier Circuits: (a) Simple Single Amplifier and (b) Two Followers and Adjustable Gain Added to Circuit in (a)

There are several other special amplifiers which are useful in conditioning the input signal in a measuring system. An *isolation amplifier* is used to amplify a differential signal which is superimposed on a very high common-mode voltage, perhaps several hundred or even several thousand volts. The isolation amplifier has the characteristics of an instrumentation amplifier with a very high common-mode voltage capability. Another special amplifier, the *chopper stabilized amplifier*, is used to accurately amplify microvolt level signals to the required amplitude. This amplifier employs a special switching stabilizer which gives extremely low input offset voltage drift. Another useful device, the *electrometer amplifier*, has ultra-low input bias currents (generally less than one picoampere) and is used to convert extremely small signal currents into a high-level voltage.

Below, some additional definitions are given that apply to the above types of instrumentation amplifiers:

1. *Input offset current* is the difference in the bias current of the inverting and noninverting inputs of an operational amplifier.
2. *Input offset voltage* is small, undesired signals generated by the amplifier, appearing between the input terminals and causing some voltage output with no input voltage. The primary source of input offset voltage is emitter-based voltage mismatch of the differential input transistor in bipolar transistor amplifiers, and gate-source voltage mismatch in field effect (FET) input amplifiers.
3. *Amplifier nonlinearity* is the gain deviation from the straight line on a plot of amplifier output versus input (the transfer curve).
4. *Settling time* is the time required for output voltage to stabilize within some tolerance of its final value for a full-scale input step signal.
5. *Slew rate* is the maximum rate of change of the output voltage while the amplifier is operating.
6. *Common-mode rejection (CMR)* is a measure of how output voltage changes when both inputs change by equal amounts. This specification usually is given for full-range input voltage change and a specified source impedance imbalance.

7.2.2 Common-Mode Rejection[3]

For measuring low-level signals from remote sources, adequate common-mode rejection is required. An analytical model for calculating the *common-mode rejection rate (CMRR)* is shown in Fig. 7.2a, where: A_1 and A_2 are the gain magnitudes between the output and inputs 1 and 2, Z_{O1} and Z_{O2} and "common-mode" input impedances at terminals 1 and 2, Z_{12} is "differential" input impedance between terminals 1 and 2, Z_1 and Z_2 are source impedances, C_1 and C_2 are capacitances from the inputs to ground, including input capacitance of the device itself, e_s is the differential input signal which is equal to $e_1 - e_2$, and CMV is the common-mode voltage which is equal to $e_1 - e_s$.

The common-mode rejection ratio (CMRR) is the ratio of the CMV to the contribution of the output due to CMV alone, i.e., CMV/Δe_{out}, where Δe_{out} is referred to the input. This parameter is expressed in dB, as follows:

$$\text{CMRR} = 20 \log \left[\frac{\text{CMV}}{\Delta e_{out}} \right] \quad (7.3)$$

For the device alone, then:

$$\text{CMRR} = 20 \log \left[\frac{1}{A_1 - A_2} \right] \quad (7.4)$$

Since A_1 and A_2 are inherently frequency-dependent and may be amplitude-dependent, the CMRR behavior of a device such as the amplifier in Fig. 7.2a is a function of the frequency and sometimes the magnitude of CMV. The *effective* CMRR of the complete circuit is dependent on both the external and internal impedance balances.

In most cases, the impedances Z_{O1} and Z_{O2} are high enough with respect to Z_1 and Z_2, and the reactances of C_1 and C_2 can be omitted. The impedance Z_{12} is also usually high enough to prevent significant source loading. Under these conditions, the external circuit imbalances, including the input capacitances of the device, can cause a significant reduction in overall CMRR. To calculate effective external CMRR, Fig. 7.2b can be used, where:

$$\frac{1}{\omega C_1} \gg R_1, \frac{1}{\omega C_2} \gg R_2, R_1 - R_2 = \Delta R, \text{ and } C_1 - C_2 = \Delta C \quad (7.5)$$

Measurement/Automation

then:

$$\text{CMRR} \cong 20 \log \left[\frac{1}{\omega(R_1C_1 - R_2C_2)} \right] \quad (7.6)$$

$$\text{CMRR} \cong 20 \log \left[\frac{1}{\omega(R_1\Delta C + C_2\Delta R)} \right] \quad (7.7)$$

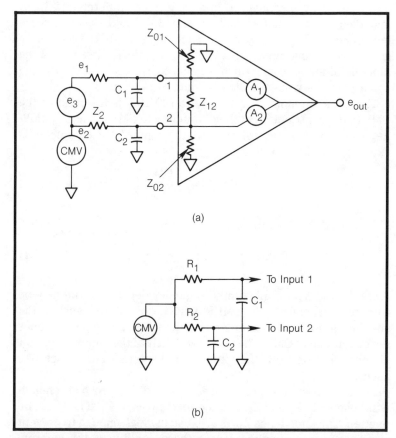

Figure 7.2—(a) Model for CMRR Calculation and (b) External Circuitry for Calculation of Effective CMRR

7.2.3 Isolation Amplifiers for Effective Measurements[3-7]

The *isolation amplifier* is a data or instrumentation amplifier in which the output is completely isolated from the input circuit. Isolation amplifiers help designers solve problems in instrumentation, industrial monitoring, process control and patient monitoring. These devices allow small (millivolt) signals to be processed accurately despite hundreds of volts of CMV, and safely despite CMV or fault voltages in the thousands of volts.

Because no input return path is required from the signal source to power ground, ground wires and ground loop problems may be avoided. Optically coupled and transformer coupled devices comprise presently available isolation amplifiers. When low-level signals are close to lines operating at high energy levels, there is a possibility that induced voltages will mask the low-level signals and saturate the amplifier. An amplifier with high CMV and CMR capability plus isolation can be very helpful in such applications.

Figure 7.3 shows how an isolation amplifier eliminates grounding problems in two ways. In Fig. 7.3a, since a bias-current return to the supply is not necessary, the input circuit can be independent of system ground with its logic spikes, voltage drops, etc. In Fig. 7.3b, especially with three-port isolation, the amplifier's power ground is independent of the output *low*; returning the amplifiers power ground to system ground does not cause grounding errors in the output signal line. To increase an isolation amplifier's common-mode rejection capability, stray capacitance must be reduced. As shown in Fig. 7.4, where the AD2989 isolation amplifier (from *Analog Devices, Inc.*) is used, this can be done by tying the transducer's shield to the low-input terminal. In some amplifiers, the shield is connected to a guard terminal rather than to the low input.

Measurement/Automation

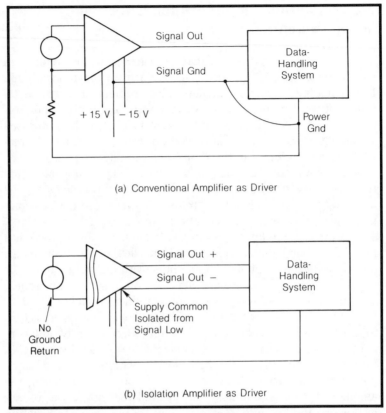

Figure 7.3—Comparison of Grounding Techniques for Conventional and Isolation Amplifiers

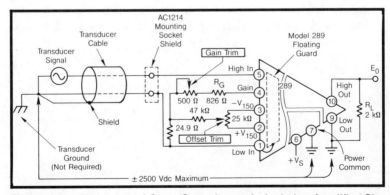

Figure 7.4—Reduction of Stray Capacitance in Isolation Amplifier Circuit (Courtesy of Analog Devices, Inc.)

7.3 Testability and Test Fixtures

Incorporating testability into printed circuit (PC) boards during the design stage helps diagnosing and isolating faults automatically. Additionally, it can reduce repair costs drastically. On the other hand, the design of test fixtures, which provide the interface between a functional digital board tester and a PC board under test, can be as critical to the success of any test project as the proper selection and use of measuring instruments.

7.3.1 Built-In Testability[8-10]

Usually, *test points* and *monitor points* are incorporated into PC boards during the design stage. Test points monitor logic values of certain circuit lines, whereas control points are driven to produce a failure mode by establishing known states. Control points can also serve as test points. Flip-flop counters, shifter-registers and other memory elements usually start at unknown states when power is applied, which makes it difficult to detect a failure. To alleviate this problem, initialization of these elements to a known condition at the start of testing is necessary. Ideally, logic elements should be reset from the external pins of the board where, in certain cases, additional logic may be required. If there are no external pins on the board, a powerup reset must be added as shown in Fig. 7.5.

In general, testability interfaces can be simplified and pitfalls avoided by following certain guidelines:
1. Divide complex logic functions into smaller, combinatorial logic segments.
2. Route text/control points to the edge connector to enable monitoring and driving of internal board functions and to assist in fault diagnosis.
3. Use a single, large-edge connector to provide I/O pins and test/control points.
4. Provide adequate decoupling at the board edge and locally at each IC.
5. Provide signals leaving the board with maximum fan-out drive, or buffer them.
6. Buffer edge-sensitive components from the edge connector—such as clock lines and flip-flop outputs.

Measurement/Automation

7. Single-load each signal entering the board whenever possible.
8. Terminate unused logic pins with a resistive pull-up to minimize noise pickup.
9. Buffer flip-flop output signals before they leave the board.
10. Avoid wire-OR and wire-AND connection.
11. Use open-collector devices with pull-up resistors to enable external override control.
12. Make PC-board I/O signals TTL-compatible to keep automatic test equipment (ATE) interface costs low and give flexibility. Investigate new devices that might provide better capabilities.

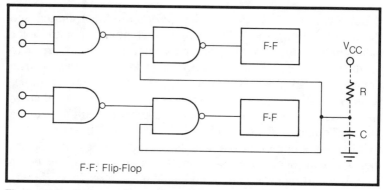

Figure 7.5—Power-up Reset Circuit (Dotted Line) Added to Provide Internal Initialization

7.3.2 Test Fixtures and Interference Problems[10-13]

In a test setup, if an interface fixture exhibits ground noise, crosstalk or impedance mismatch, the test program can suffer from intermittent failures and distorted results. Universal interface fixtures that are adaptable to all varieties of printed circuit boards do not exist. Therefore, users must undertake the task to design their own in the shop or under contract to fill their needs. Interface fixtures take several forms, but they may be generally defined as an assembly of wires, cables and connectors that connect a UUT to the ATE or tester. Two major types of fixtures are *edge-connector* and *bed-of-nails*. Either type of fixture may be *dedicated* to one type of PC board or *universal*, i.e., able to accommodate several PC board types. The latter often includes a load board that adapts the PC board to the fixture.

An edge-connector fixture with one-to-one direct wiring between the tester I/O pins and the edge connector for the UUT is the simplest and easiest to make. An intermediate connector, usually a *zero-insertion force (ZIF)* type, may be used between the UUT edge-connector and the tester connectors. Such an arrangement helps if the direct connection to the tester is difficult or if the fixture must be changed quickly from one PC family to another. A bed-of-nails fixture has small probe points that contact the PC board on the noncomponent side and usually includes a vacuum pump that sucks the board onto the probe points to maintain contact pressure.

Crosstalk and ground noise are the most common types of fixture noise. Proper fixture design can minimize crosstalk-induced noise. There are several approaches that can be used, and the most appropriate choice depends on the UUT logic switching speed, tester driver rise time and the total wiring length of the fixture. The choices include: straight wiring, flat cable with and without alternate grounds and a ground shield backing, discrete twisted pair, flat cables of twisted pairs and coaxial cables.

Straight wiring is most susceptible to crosstalk, and for lengths greater than 7.5 to 10 cm it is unacceptable. A *flat cable without alternate grounds* is a common, but also unacceptable, method for lengths more than 7.5 to 10 cm. A *flat cable with alternate grounds* is a better wiring method since ground wires between signal runs help shield one signal from another. A flat cable with alternate grounds and a ground shield backing can be even better. A *discrete twisted pair*, one pair of wires per signal, is probably the most common type of fixture wiring. *Flat cables of twisted pairs* present a neat way to handle the mechanical layout of the fixture wires, but they often exhibit worse crosstalk characteristics than bundles of discrete twisted pairs. Finally, *coaxial cables* have negligible crosstalk and are the best wiring for electrical reliability. Unfortunately, they are difficult to work with: bulky and expensive. On the plus side though, the savings from reduced reliability problems later on can make up for the higher initial cost. They are also highly recommended for realtime testers.

When wiring a fixture, a common mistake is to tie the twisted pair, or the coax cable ground, together at a single point and then route them to a UUT connector ground using a single wire. Figure 7.6 shows both incorrect and correct grounding of an edge-connector fixture. An intermediate ZIF connector, whether used for a bed-of-nails or an edge-connector, should have a solid ground bus on both sides. At least one connector pin should be dedicated to ground for every ten signal pins.

When a large number of PC board types is expected to be used in a common edge-connector fixture, a setup can be made for measuring worst-case noise; this setup can also prove helpful in a measuring noise up to ZIF connector in a bed-of-nails fixture. Figure 7.7 is a partial schematic of a test module for measuring worst-case crosstalk and ground noise. The circuitry is placed on a blank PC board that substitutes for the UUT, and Schottky open-collector logic drivers switch fixture lines from high to low with the fastest fall time possible. The tester monitors the noise on an

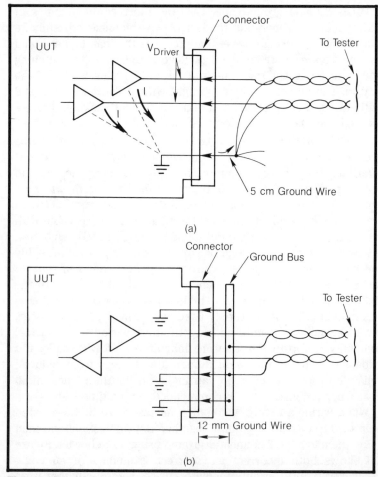

Figure 7.6—(a) Correct and (b) Incorrect Grounding of Twisted Pairs in an Edge-Connector Fixture (Adapted from Ref. 10)

Test Fixtures

undriven victim line. The flip-flop for each channel holds a "1" if the UUT is to drive the channel and an "0" if it is to be monitored. Two of the tester channels are dedicated to board inputs that can clock the flip-flops and switch the driven channels.

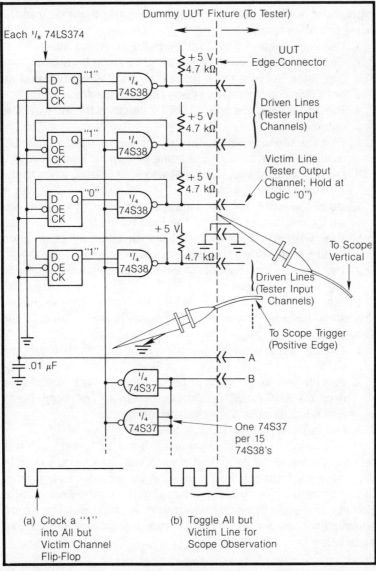

Figure 7.7—Noise Test Module for Measuring Worst-Case Crosstalk and Ground Noise and obtaining a Noise Profile of the Test Fixture (Reproduced from *Electronics Test* by Permission of Morgan-Grampian Publishing, Co. and Author N.D. Megill)

7.3.3 Guidelines for Fixture Design and Interconnections[13-15]

Both crosstalk and ground problems can be reduced in a digital test fixture by following certain empirical guidelines. Design guidelines, for example, for edge-connector fixtures may be summarized as follows:
1. Use coax cables or discrete pairs for all signal runs.
2. Connect all coax or twisted-pair grounds at both ends.
3. Employ a heavy ground bus at the UUT connector and on both sides of any intermediate (ZIF) connector.
4. Provide at least one ground per 10 signals in the intermediate connector.
5. Provide a load board connector for customizing a universal fixture and use a good grounding system on load boards.
6. Use ferrite beads on the signal wires, or twisted pairs, at the UUT end of the fixture.
7. Do not mount noise filtering capacitors directly on a universal fixture; mount them on the load board.
8. Use twisted pairs or flat cables with alternate grounds on fixture accessories. Ground them at both ends.
9. Use terminating networks to reduce reflections and decouple any V_{cc} connections.

Design guidelines for bed-of-nail fixtures to include the general guidelines for edge-connector fixtures, with the addition of the following:
1. Provide a copper-etch ground plane directly beneath the UUT,
2. Provide at least 12 ground probe points, evenly distributed over the UUT surface. Connect them to the copper-etch ground plane with very short wires and
3. Use twisted pairs in the test head.

On the other hand, implementing basic D/A and A/D test techniques requires a good deal of care, skill, and feedthrough. In the converter industry, it is important to separate analog from digital activities by physically separating the connections as size permits, and often providing separate grounds. The following are some guidelines for proper converter test fixture design and installation:

1. Separate analog grounds from all other system, power and digital grounds. Make sure the analog ground does not find its way back to another ground deep within the circuitry (see also subsection 6.4.3).
2. Form large ground etches. If wire is employed, care should be taken to keep it away from EMI sources, such as modular power supplies, which can radiate 60 or 50 Hz into both ground and small signal circuits, causing substantial noise.
3. Use decoupling as close as practicable to the converter under test; do not tie the decoupling capacitors to the analog ground.
4. Investigate the source impedance of the analog source, since an abnormally high source impedance can severely hinder performance. For 14-bit and higher D/A converters, the input impedance of the measurement system proves critical. To compare the precision divider to the device under test (DUT), a high-impedance null meter must be used.

7.4 Analyzers and Automatic Test Interfaces

Dedicated logic analyzers minimize setup problems and reduce fault finding to simple go, no-go comparisons. On the other hand, automated RFI and EMI measurement systems, using spectrum analyzers and desktop computers, represent a leap in convenience and performance over manual measurements. In automatic testing the acceptance of GPIB has resulted in simplification and significant reduction of the engineering costs of putting systems together.

7.4.1 Dedicated Logic Analyzers

Logic analyzers have three important advantages over traditional methods of digital troubleshooting. Perhaps the most significant benefit is the ability to record multiple channels simultaneously. Logic analyzers also can easily trigger upon a unique selected condition (address or data word or a combination of the two) among the many channels monitored, and the same trigger can be used to stop the data-collection process as well. These features allow the examination of multichannel events to occur both before and after a system crash and, if desired, trigger the crash itself.

Both logic-state and logic-timing analyzers have developed as general-purpose instruments, but the increasing complexity of the systems they serve has led to growing setup problems for the user. The problems inherent with both types of analyzers can be reduced by the use of a dedicated instrument. Such an analyzer is made possible by the bus-oriented design of many modern systems. Designed as a single PC board, it can be plugged into any slot in a bus-oriented system. Thus, there are no probes to connect; address, data and qualifier information are fed to the analyzer through the edge connector by the bus itself. The number of signals monitored is limited only by the number of pinouts on the card edge. For example, by using the popular S-100 computer bus, up to 100 pinouts can be monitored by a dedicated logic analyzer.[16]

7.4.2 Receivers and Spectrum Analyzers for RFI/EMI Measurements

Two major instrument types can serve EMI-measurement tasks: the *receiver-type* device and the *spectrum analyzer*. Both are *superheterodyne receivers* capable of tuning over a wide range. The receiver type instrument—optimized for sensitivity and spurious-signal-rejection capability—provides RF stage and usually a variety of detector functions.

A receiver able to detect and measure RFI/EMI, and relate to precise (accepted) international standards must necessarily be based upon a different design than that of a communications receiver. The latter type normally covers 10 kHz to 1,000 MHz frequency range and is primarily designed to receive and decode selected transmission. Conversely, an EMI receiver must include these design characteristics:[17] a greater instantaneous dynamic range than that of a communications receiver, a circuit that monitors the maximum allowable voltage at different stages to prevent short-term overload, detector time constants that conform to internally agreed-upon standards appropriate IF bandwidth and precise amplitude calibration over the operating frequency range. If active antennas are used to measure the field in the low-frequency range, they must have the necessary dynamic range and proper antenna correction factor.

The spectrum analyzer, on the other hand, is optimized for automatic sweeping over a wide frequency range and for use with a video display. It contains an untuned broadband-mixer front end. Also, because of the nature of a swept video display, it usually provides only a peak detector. Despite the receiver type's intrinsically higher sensitivity and rejection, the sensitivity and spurious rejection capability of the spectrum analyzer can be boosted by using preamplifiers, filters or tracking preselectors. Detector functions, such as quasi-peak response, can also be added.[18]

The spectrum analyzer is essential for efficient and thorough EMI testing. The recently introduced Tektronix 494/494P spectrum analyzers are examples of these instruments. Both operate from 10 kHz to 21 GHz (or to 325 GHz with external waveguide mixers) and feature a minimum resolution bandwidth of 30 Hz at frequencies of up to 60 GHz.[19] Indoor and outdoor EMI tests can also be performed with Hewlett-Packard's HP8568B or 8566B spectrum analyzer/EMI receiver systems. These analyzers, combined with other equipment, probes, and adapters, can form a system able to locate sources of problem emissions and evaluate potential fixes (Fig. 7.8).

Figure 7.8—Spectrum Analyzer/EMI Receiver System that automates FCC, VDE, and MIL-STD Emission Measurements. (Courtesy of Hewlett-Packard Company).

7.4.3 GPIB-Based Automated Systems[20-22]

The *general purpose interface bus (GPIB/IEEE 488)* standard for instrumentation has resulted in simplification of the design of automated test systems. Low-cost GPIB controllers and simple program codes of current instruments allow easy transition from manual operator control of the instrument's functions, ranges and output levels. In order to properly automate these tasks, the possible switching needs of the test setups that will be automated must first be defined. Table 7.1 summarizes the most important characteristic requirements for switching performance. It must be noted that several of the categories shown in Table 7.1 can be combined since they share common characteristics. A typical automatic test system using GPIB controller is shown in Fig. 7.9. All of the instruments are connected to the data bus by GPIB cables. The cables may be connected in a star or linear fashion, as the construction of the connectors permits stacking. All data outputs and programming inputs use a single connector at each device to pass to or from the bus.

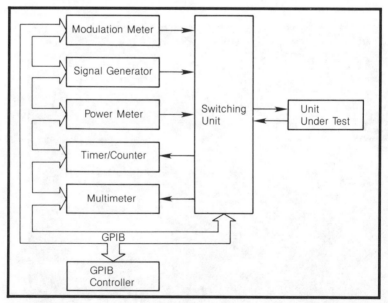

Figure 7.9—A Typical Automatic Test System Using GPHB Controller

Table 7.1 Characteristic Requirements for Switching Performance (After Ref. 20)

Category	Max Voltage	Maximum Current	Offset	Frequency Range	Circuit Impedance
Low level	<50 V	<100 mA	<1 μV	DC–100 Hz	High
Audio	>50 V	<5A	<1 mV	20 Hz–50 kHz	4–600 Ohms
High current	<30 V	>10 A	<1 mV	DC	Low (<1 Ohm)
High voltage	>500 V	<100 mA	<100 uV	DC–20 kHz	High
TTL sourcing	5 V	>8 mA	N/A	N/A	N/A
Relay driver	>24 V	>20 mA	N/A	N/A	Low (<500 Ohms) inductive
Power line	>265 V RMS	>1 A	N/A	45–440 Hz	Low
RMS AC	>300 V RMS	<100 mA	100 μV	10 Hz–10 MHz	High or low
RF	<30 V peak	<100 mA	<1 mV	DC–100 MHz	50 or 75 Ohms
VHF/UHF	<50 V RMS	>30 W >1 A	<1 mV	10 MHz–1 GHz	50 or 75 Ohms
Fast pulse	<10 V peak	<200 mA	<100 μV	DC–1 GHz*	50 Ohm
Microwave	<50 V RMS	>30 W >1 A	N/A	10 MHz–18 GHz	50 Ohm
Logic signals	<7 V peak	<100 mA	<1 mV	DC–100 MHz	Low

*Effective bandwidth.

7.5 FCC Rules and Digital Equipment: Testing for Compliance

Unwanted signals might be emitted from an instrument or a system at the same time that a useful signal is being emitted. This source could be termed the *aggressor*. Alternatively, the instrument or the system could be susceptible to interference from an outside source and is, therefore, the *victim*. Since these unwanted signals can be either radiated or conducted, there are basically four types of EMI measurements: conducted emission, radiated emission, conducted susceptibility and radiated susceptibility.[23] Knowing these emission and susceptibility problems and understanding EMI test methods eases product acceptance.

7.5.1 FCC Computing Device Rules

A computing device is briefly defined as any electronic device that uses digital technology and generates RF energy above 10 kHz. Examples of such devices include, but are not limited to: personal computers, calculators, data terminals, modems, telephones and data processing equipment. The various peripherals, whether considered separately or not, are subject to the same limits as the host computer. As of October 1, 1983, computing devices manufactured after that date are subject to the computing devices rules.

The *FCC rules* established two device classifications. The first group, *Class A devices*, includes computing products intended for use in commercial, industrial and business environments. *Class B devices* comprise computing equipment, such as personal computers and peripherals and all types of electronic games, calculators and watches that are termed residential equipment. Table 7.2 lists the EMI requirements for both classes of equipment along with some other pertinent information.[24]

FCC Standard

Table 7.2 Computing Device Requirements (47 CFR 15.801-15.840)*

- Class A—Commercial
- Class B—Residential
- Exemptions: Home appliances
    Medical devices
    Test equipment
    Equipment used exclusively in industrial plant
    Equipment in vehicles
- Radiated limit: (MP-4)

Freq (MHz)	Class A (μV/m at 30 m)	Class B (μV/m at 3 m)
30–88	30	100
88–216	50	150
216–1000	70	200

- Conducted limit: (MP-4)

0.45–1.6	1000	250
1.6–30	3000	250

- Only personal computer and peripherals certificated, all only other devices verified by manufacturer
- Label and user information required

*After Ref. 24.

7.5.2 The FCC Standard of Measurement of Radio Noise Emissions from Computing Devices[25]

This standard covers uniform methods of measurement of radio noise emitted from computing devices defined in section 15.4 of FCC Rules. The technical standards for computing devices are set forth in Subpart J of Part 15 of FCC Rules (47 CFR Part 15J). Methods which are covered include measurement of radiated and power line conducted radio noise.

In using the standard definitions in Parts 2 and 15 of FCC Rules, the following definitions apply:

1. *Equipment Under Test (EUT)* is a representative computing device or system, peripheral, etc., being tested or evaluated.
2. *Ambient Level* is the magnitude of radiated or conducted signals and noise existing at a specific test location and time.
3. *Emission* is electromagnetic energy produced by a device which is radiated into space or conducted along wires, and is capable of being measured.

Measurement/Automation

4. *Ground Plane* is a conducting surface used to provide uniform reflection of an impinging electromagnetic wave. Also, the common reference point for electrical potentials.
5. *Line-Impedance Stabilization Network (LISN)* is a network (sometimes called mains network) inserted in the supply mains lead of the EUT, which provides a specified measuring impedance for radio-noise voltage measurement and isolates the EUT and the measuring equipment from the supply mains at radio frequencies.
6. *Radio Noise* is electromagnetic energy in the radio frequency range that may be superimposed upon a wanted signal.
7. *Radiated Radio Noise* is radio noise radiated into space. Such noise may include both the radiation and induction components of the field.
8. *Random Noise* is electromagnetic disturbance (noise) originating in a large number of discrete disturbances with random occurrence in time and amplitude. The term is most frequently applied to the limiting case where the number of transient disturbances per unit time is large, so that the spectral characteristics are the same as those of thermal noise.
9. *Narrowband Radio Noise* is radio noise having a spectrum exhibiting one or more sharp peaks, narrow in width compared to the nominal bandwidth of the measuring instrument, and far enough apart in frequency to be resolvable by the instrument.
10. *Broadband Radio Noise* is radio noise having a spectrum broad in width, as compared to the nominal bandwidth of the measuring instrument, and whose spectral components are sufficiently close together in frequency and uniformity that the measuring instrument cannot resolve them.

7.5.3 Conducted-Emission Tests[25-27]

Evaluating the conducted emissions from a piece of equipment requires a virtually RFI-free environment. Power lines are heavily filtered, but sources that might radiate energy are isolated, usually in a shielded room. Furthermore, the measuring instruments used in that room must be noise-free below specified maximum levels.

Conducted-Emission Tests

A *line impedance stabilization network (LISN)*, shown in Fig. 7.10, provides a known line impedance over the frequency range of interest. For this network, it is important to note that care should be exercised in the design and construction of the inductive and capacitive elements to ensure that the desired impedance characteristic is met throughout the frequency band of interest and at the current of rating of the equipment to be tested.

As previously mentioned, a *spectrum analyzer* that measures the voltage (usually the peak voltage) over the spectrum range is fast and convenient. The *quasi-peak* detectors found in tuned radio receivers can also measure the interference value of noise signals as specified by the FCC. In essence, a quasi-peak circuit is a peak signal detector followed by an integrator so that occasional peaks of noise are not at their peak value, but recorded by the time constant of the integration circuit (Fig. 7.11). Therefore, these detectors should be limited to checking the effects on analog radio communications equipment—not digital devices.

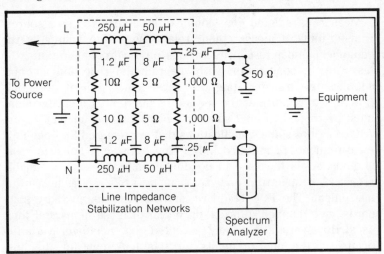

Figure 7.10—Conductive RFI Emission Testing

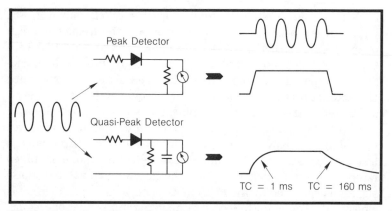

Figure 7.11—Peak/Quasi-Peak Detector

7.5.4 Radiated-Emission Tests

Measuring radiated emissions can be considered as a source/transfer-function/measurement relationship. The source is the equipment under test in its various physical configurations; the test site, in conjunction with the radiation-receiving antenna, supplies the transfer function. These measurement potentials (microvolts region), at the measuring-instrument input terminal, must be related to field strength (microvolts per meter). Figure 7.12 describes some specifications that are required for open-field measurements of radiated EMI.[19] The transducer consists of an antenna. As is the case in conducted-emission tests, a tunable receiver, or spectrum analyzer, can be used as the measuring instrument. The FCC standard, and the various auxiliary standards, give detailed information on all of these items and their uses. However, it is widely accepted that a greater variation occurs in the measurements for this test than for the line-conducted-emissions tests.[18]

Radiated-Emission Tests

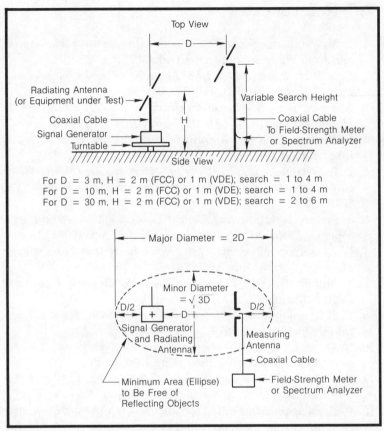

Figure 7.12—Test Setup for Open-Field Radiated EMI Measurements (Adapted from Ref. 19)

7.6 References

1. L. M. Faulkenbery, *An Introduction to Operational Amplifiers*, John Wiley & Sons, New York, 1977.
2. E. L. Zuch, *Principles of Data Acquisition and Conversion—Part 2, Digital Design*, July 1979, p. 30.
3. Analogic, *Design & Selection Guide to A/D & D/A Conversion Modules, Bulletin No. ADASFC INT. REV. 3*, 1979, p. 5.
4. F. Pouliot, *Isolation Amplifiers for Effective Data Acquisition, Analog Dialog*, Vol. 9, No. 2, 1975, pp. 10–11.
5. P. H. Garret, *Analog Systems for Microprocessors and Minicomputers*, Reston Publishing Company, Inc., Reston, VA, 1978.
6. J. Sylvan, *Isolation and Conditioning Clean Up Industrial Signals, Electronic Design*, April 29, 1982, pp. 169–172.
7. C. J. Georgopoulos, *Squelching Noise in Instrument Systems, Machine Design*, July 22, 1971, pp. 74–79.
8. J. Mittelbach, *Put Testability Into PC Boards, Electronic Design 12*, June 7, 1978, pp. 128–131.
9. C. J. Georgopoulos, *Self-Testing in Automatic Testers, Digital Design*, January 1973, pp. 38–44.
10. N. D. Megill, *Test Fixture Noise, Part 1: Symptoms, Causes and Cures, Electronics Tests*, May 1980, p. 49.
11. R. E. Tulloss, *Automated Board Testing: Coping With Complex Circuits, IEEE Spectrum*, July 1983, pp. 38–43.
12. P. K. Lala, *Onchip Fault Tolerant Design Scheme, Computer Design*, August 1983, pp. 143–146.
13. N. D. Megill, *Test Fixture Noise, Part 2: Pinpointing and Curing Common Errors, Electronics Tests*, June 1980, p. 30.
14. E. B. Sinclair, *Floating Measurements Need Not Be Dangerous, Electronic Design*, June 11, 1981, pp. 235–238.
15. J. Lang, *Testing Digital and Analog Converters: A Technical Update, Electronics Test*, June 1980, p. 40.
16. I. H. Spector, *Dedicated Logic Analyzer Minimizes Setup Problems, Electronics*, September 28, 1978, pp. 136–139.
17. U. Rohde, *EMI/RFI Test Receivers, Ham Radio*, November 1983, p. 70.
18. I. Straus, *Testing Products Correctly Ensures EMI-Spec Compliance, EDN*, November 25, 1981, pp. 121–130.
19. H. Benitez, *Use a Spectrum Analyzer to Make EMI Measurements, Microwaves & RF*, April 1984, p. 134.

20. N. Laengrich, *New Solutions of Switching and Timing Requirements of GPIB-Based Systems, Racal-Dana GPIB Test Systems Handbook*, Publication, No. 980593.
21. N. Laengrich, *IEEE 488 Short-Run Testing, Electronics Test*, December 1982, p. 71.
22. A. W. Conway, *Building a Compact Test Instrumentation System, Racal-Dana GPIB Test Systems Handbook*, Publication No. 980593.
23. G. Sorger, *Measuring RFI—Spectrum Analyzer vs RFI Receiver, Communications Engineering International*, April 1984, p. 53.
24. R. Fabina and J. Husnay, *FCC Experience in Measuring Computing Devices*, IEEE 1985 International Symposium on Electromagnetic Compatibility, Wakefield, MA, August 20–22, 1985.
25. Office of Science and Technology, *FCC Methods of Measurement of Radio Noise Emissions from Computing Devices*, Federal Communications Commission, Washington, D.C., FCC/OST MP-4, 1983.
26. W. T. Mitchel, *FCC and VDE Impose Tight Conduct RFI/EMI Specs, Electronic Design*, December 23, 1982, pp. 141–147.
27. M. B. Head, *RFI Power Line Filters: Selection, Specification and Testing*, Midcon/81, Chicago, IL, November 10–12, 1981.

General References

- A. Jenkins and D. Bowers, *Instrumentation AMP IC Delivers High Current, Electronic Design*, June 9, 1983, p. 171.
- J. Graeme, *Instrumentation Amplifiers Shift Signals from Noise, Electronic Design*, September 13, 1980, pp. 119–123.
- R. Morrison, *Grounding in Instrumentation Systems, EMC Technology*, January-March 1983, p. 50.
- K. Rush, *Understand Probe Impedance to Assess Signal Distortions, EDN*, April 18, 1985. pp. 247–252.
- D. W. Raymond, *In-Circuit Testing Comes of Age, Computer Design*, August 1981, p. 117.
- R. Allan, *ATE Lays the Cornerstone of the Automated Factory, Electronic Design*, April 5, 1984, pp. 84–91.
- N. Urdaneta, *Timing Considerations of GPIB Systems*, Wescon/80, Anaheim, CA, September 16–18, 1980.

Measurement/Automation

- R. W. Johnson, *FCC Rules and Digital Equipment: Testing for Compliance, Test & Measurement World*, April 1982, pp. 16–20.
- M. Nave, *Line Impedance Stabilization Networks: Theory and Use, RF Design*, April 1985, p. 54.
- M. T. Ma, et al., *A Review of Electromagnetic Compatibility/ Interference Measurement Technologies, Proceedings of the IEEE*, Vol. 73, No. 3, March 1985, pp. 388–411.

Chapter 8
Optoelectronic Systems Interfaces

Taking advantage of the wave-propagation properties of light, optoelectronic devices offer electrical isolation, ground loop elimination, protection from electrical sparks and high immunity to electromagnetic interference. In this chapter, various interfaces of optoelectronic systems are discussed, including: optical isolators, fiber optic links and infrared transmission equipment in confined spaces.

8.1 Introduction

Optical isolators, also called *optocouplers*, play a major role as isolation elements in control systems, in telephone communications and as line receivers in digital data communications. In addition, these devices have many linear applications: for example, as differential isolation amplifiers, ac-coupled and servo-isolation amplifiers.

Fiber optics is a dynamic and rapidly expanding technology. Advantages of fiber optics over conventional conductors depend on where they are used, especially for application in industrial environments. The advantages include: immunity to ground loops, insensitivity to EMI and RFI, isolation, low transmission loss, wide bandwidth and flexibility.

Infrared (IR) systems, while preserving a high level of EMI immunity, have the further advantages of requiring no cable installation and can be used as mobile links. They offer substantial cost savings when used in confined spaces where information is transmitted via diffused IR light.

8.2 Optical Isolators

An optical isolator, also called an optical coupler or an optically coupled isolator, is a device which converts an electrical signal at its input, via a light source, into photon energy. It then optically couples this energy to a photosensor to produce a replica of the input at its output.

8.2.1 Basic Types and Applications[1]

In the optical isolator, or photon-coupled pair, the coupling is achieved by light being generated on one side of a transparent insulating gap and then being detected on the other side of the gap without an electrical connection between the two sides (except for a coupling capacitance of approximately 1 pF). Modern semiconductor optical isolators use the *light emitting diode (LED)* as a light source. The LED is usually made of gallium-arsenide, operates in the near-infrared region and is optically coupled to a photo-sensing junction. This junction may be part of a photodiode, a photo-transistor or a photo-Silicon Controlled Rectifier (SCR), as shown in Fig. 8.1.

Optical isolators have a host of applications. These include:
1. Isolating different voltage levels between circuits,
2. Preventing interference between control and power circuits, using the unidirectional feature,
3. Insulating people or low-voltage circuits from the hazards of high voltage shock.
4. Eliminating dc ground loops,
5. Amplifying or attenuating signals,
6. Performing on/off switching and
7. Interfacing microprocessor-based systems to their cooperating environment.

Digital Links

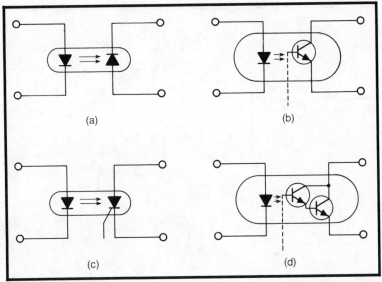

Figure 8.1—Basic Types of Optical Isolators: (a) LED-Photodiode, (b) Phototransistor with or without Base Terminal (c) LED-Photo-SCR and (d) LED-Photo-Darlington

8.2.2 Digital Links With Optical Isolators

By using an optical isolator between two systems coupled by a transmission line, effective line isolation can be achieved. Figure 8.2a shows a typical interface system using TTL integrated circuitry coupled by a twisted pair. Optical couplers can also effectively isolate interface circuitry to a microcomputer or microprocessor-based digital system as shown in Fig. 8.2b.

Isolation transformers or relays could accomplish the above tasks, but they would not be as fast as optical isolators. Also, line drivers and line receivers could be used to eliminate the noise and increase the speed, but they would be ineffective if there were high potential differences between the input and output. Digital input units containing optical isolators protect the microprocessor or microcomputer from standard as well as overvoltage conditions.

Optoelectronic Interfaces

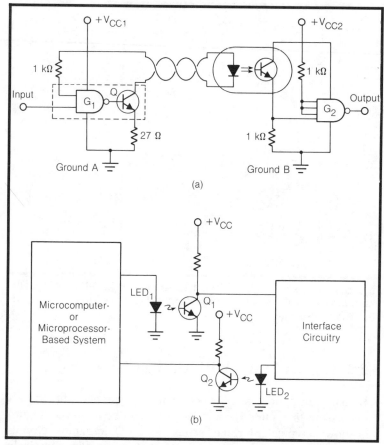

Figure 8.2—Digital Interface Isolation Using Optical Isolators in (a) Unidirectional and (b) Bidirectional Communications

8.2.3 Linear Operation[1-4]

It is perhaps not as well-known that optical isolators are also suitable for analog circuits. With the correct circuit techniques, the isolator's advantages can be applied to such linear tasks as sensing circuits, patient-monitoring equipment, adaptive controls, power supplies and audio or video amplifiers.

Using coupler pairs with good tracking in their transfer function, it is possible to eliminate the nonlinearity effects of the optical isolators (optocouplers). A circuit for linear signal transmission that consists of a pair of optical isolators in a feedback

Linear Operation

arrangement is shown in Fig. 8.3 It includes a matched pair of LED-photodiode optical isolators that operate in the feedback path of an operational amplifier to provide linear signal transmission with stable gain and high-voltage protection. The advantage of this circuit is that the operational amplifier is also protected against high voltage.

Considering a certain range of input current, the transfer function can be approximated as follows:[3]

$$I_{PD} = C(1 + m)^n \tag{8.1}$$

where, $C = I_{LED}(dc) \times CTR$
dc current-transfer ratio $CTR = I_{PD}(dc)/I_{LED}(dc)$
modulation index $m = I_{LED_p}(ac)/I_{LED}(dc)$

Using Eq. (8.1), an expression can be written for the output voltage of the circuit of Fig. 8.3:

$$V_{OUT}(t) = V_{OUT_{dc}} \{C_1[1+m_1(t)]^{n_1} - C_2[1+m_2(t)]^{n_2}\} \times R \times G \tag{8.2}$$

where, $V_{OUT}(t) = V_{OUT_{dc}} + v_{OUT}$ and $m_1(t), m_2(t)$ are input, output modulation indices, respectively.

With $G \to \infty$ and $V_{IN}(t) = V_{IN_{dc}} + V_{IN}$, Eq. (8.2) becomes:

$$V_{OUT}(t) = V_{OUT_{dc}} \times \left(\frac{C_1}{C_2}\right)^{1/n_2} \times \left\{\frac{V_{IN}(t)}{V_{IN_{dc}}}\right\}^{n_1/n_2} \tag{8.3}$$

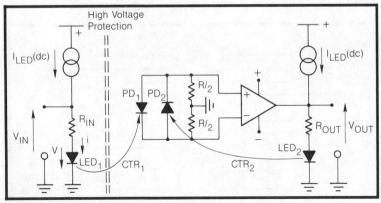

Figure 8.3—Linear Optocoupler Circuit (CTR = Current Transfer Ratio and G = Gain)[3]

Optoelectronic Interfaces

8.2.4 Example of Laser Power Supply Isolation

In a *gas discharge laser* system, the primary power source may be either the power mains, dc voltages generated within the system or the batteries. Converting the energy available from these sources into the unusual starting and running voltages and currents required by the gas discharge laser has necessitated the development of a unique *laser power supply (LPS)*. The LPS has progressed from early 60 Hz input linear designs, which were large and inefficient, to modern high-frequency (25 kHz) switching supplies, which offer mechanical and electrical performance improvements of a kind previously unattainable.

Laser power supplies, using ac mains as the source, typically have the input power galvanically isolated from the laser tube output and earth ground. This makes it possible to put the laser tube cathode at earth ground potential, with an arc path to a ground of lower impedance. The standard power supply, shown in Fig. 8.4, is isolated at two points, the power transformer and the feedback transformer. Special remote enables, on shutdown units, are isolated also by an optocoupler.[5] It is possible to have sufficient insulation thickness, and creepage distance, so that the transformers will pass a 2,500 V_{dc} hipot test. However, ac isolation is far more complicated than dc isolation. Even though there is no direct conductor connecting the primary and secondary, there are electric fields that exist between them and result in an ac leakage current. The solution is to make the isolation point different from the laser power supply, which entails either using a

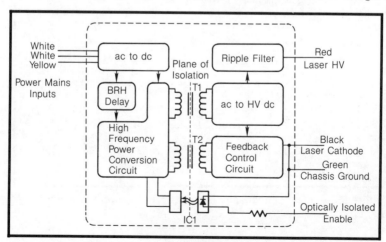

Figure 8.4—Schematic of a Standard Laser Power Supply Using Isolating Transformers and Optocoupler (Adapted from Ref. 5)

low-ac-leakage isolation transformer or switching to a dc power source with low ac leakage and utilizing a dc laser power supply.

8.3 Cables and Connectors in Fiber Optic Link

The practice of employing a fiber optic link to transfer data from one place to another is becoming common because of the wide bandwidth and immunity from external interference that this approach provides. The average designer will explore the optical and electrical characteristics of the available fiber optic systems and choose a combination that suits the needs of the application. On the other hand, if commercially available systems do not offer exactly what is wanted (or are too expensive), the designer may decide to design one. It is therefore important for the designer to be familiar with the various components involved in a fiber optic link as well as with their characteristics and compatibility in which the system is to be used.

8.3.1 Typical Fiber Optic Link

Figure 8.5 shows a typical fiber optic link between a microcomputer and a terminal device. Electrical signals, at the I/O module of the minicomputer, are converted by the light source into optical signals that are injected into the fiber optic cable that has connectors at both ends. At the I/O module of the terminal (CRT), optical signals are converted back to electrical signals by a light detector for further processing.

Fiber optic cable and connector characteristics are reviewed in this section, whereas light sources and light detectors are discussed in the next section.

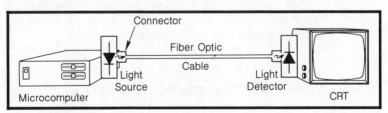

Figure 8.5—Typical Fiber Optic Link

8.3.2 Fiber Optic Cable Advantages

The advantages of fiber optic cables over metallic wires for transmitting information depend on the application, but include the following:[6]
1. Insensitivity to electromagnetic and radio frequency interference. (Note: Fiber optics may be susceptible to EMP gamma radiation unless special techniques have been used during their manufacture),
2. Immunity to ground-loop problems,
3. Improved security compared with electronic cabling (no crosstalk among parallel cables),
4. Elimination of combustion or sparks caused by short circuits,
5. Flexibility in upgrading system capacity without need to install new cables,
6. Low transmission loss (greater distance between repeaters),
7. Wide transmission bandwidth,
8. Potential low cost,
9. Relatively small size, light weight, high strength, and flexibility,
10. Suitability for digital communications and pulse modulation methods (fiber optic cable losses are independent of transmission frequency),
11. Fewer government regulatory difficulties (because of the elimination of frequency allocation) and
12. Suitability for relatively high temperature.

In the near term, optical interconnects are expected to be used primarily in point-to-point applications, where several parallel channels are multiplexed and guided by fiber optics to be demultiplexed at the receiving end and in the area of clock distribution in computers. Also, in the future it may be possible to exploit the unique properties of optical processing to perform protocol functions at the network nodes without having to convert from optics to electronics, and vice versa. This would dramatically reduce communication processing delays.[7]

8.3.3 Types and Transmission Properties

The basic element of a fiber-optic cable is the fiber itself: a thin filament of glass or plastic that transmits visible or infrared light.

It consists of two parts: the *core* and the *cladding*. Basically, light is guided along the length of the core because of reflections at the interface between the core and the cladding.

There are three basic fiber optic types: (1) multimode-step index, (2) multimode-graded index and (3) single mode-step index, as shown in Fig. 8.6.[8] They are distinguished primarily by the profile of the index of refraction across the fiber's cross-section. The *index of refraction (n)* is one of the main parameters of an optical fiber. Classically, for both glass and plastic, the index of refraction is defined as n = c/v, where c is the speed of light in vacuum and v its speed within the material. An optical fiber's *index profile* refers to the way its index varies as a function of radial distance from the fiber's center.

The simplest way to consider transmission over an optical fiber is to think in terms of total reflection in a medium of refractive index n_1 at the boundary with a medium n_2, where n_1 is greater than n_2. This is the case of a multimode-step index fiber such as shown in Fig. 8.7. Light launched into the core at angles up to θ_1

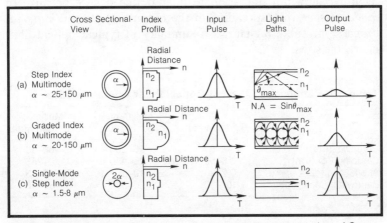

Figure 8.6—Types of Optical Fibers: Schematic Representation of Cross-Sections. Index of Refraction, Distributions and Optical Ray Paths[8]

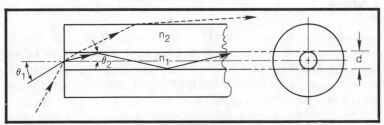

Figure 8.7—Ray Paths in a Multimode-Step Index Fiber Optic

will be propagated within the core at angles up to θ_2 to the axis. Light launched at angles greater than θ_1 (as shown by broken line in Fig. 8.7) will not be internally reflected, but will be refracted into the cladding, or possibly even out of the cladding, into the air at the second boundary if the launching angle is large enough, and n_1 and n_2 are small enough. Very pure glass fibers have yielded extremely low attenuation, in some cases much less than 1 dB/km.

Fiber optics can be grouped by wavelength characteristics as follows: *First-window* fiber passes a wavelength of approximately 850 nm. *Second window* fiber operates at approximately 1,300 nm, accounting for low loss and, as a result, longer cable runs between repeaters. *Third window* fiber is more efficient in the region of 1,550 nm. There are also *multi-window* fibers that function well at two or three wavelengths.

The bandwidth of a fiber depends on its length but greatly exceeds that of conventional coaxial cable or parallel wire; for coaxial cable, bandwidth varies inversely as the square of length and in optical fiber it varies roughly inversely by the 0.75 power of length. Bandwidth limitation in fiber optics occurs because of *modal dispersion* (spreading of the output signal) and *material dispersion*. Table 8.1 shows a summary of typical optical fiber characteristics.

Table 8.1 Summary of Typical Optical Fiber Characteristics

	Single-Mode	Multimode Step-Index	Multimode Graded-Index
Core Dia. (μm)	2–8	90–300	60–90
Cladding OD (μm)	15–50	125–450	125
Attenuation (dB/km)	0.2–2	10–50	7–15
Bandwidth (MHz-km)	10^3–10^5	20–50	150–400

8.3.4 Environmental Criteria for Fiber Optic Cables

Among the most important environmental considerations in fiber optic cable selection are[9] the range of temperatures which the cable will experience over its lifetime, potential mechanical abuse, i.e., impact, crushing, flexing tension caused by installation, wind and ice loading, potential ingress of water into the cable, and the resultant weakening of fiber strength due to stress corrosion of fiber damage due to ice formation and potential damage due to gnawing rodents.

For most of the environmental factors listed above, the materials and designs which have been successfully employed for many years in the manufacture of metallic cables are applicable to optical cables without major modifications. However, standardized specifications for optical cables are required which will assist both users and manufacturers. Some of the groups working on fiber optic standardization are: the CCITT (Study Group IV), IEC (Technical Committee No. 46E), DoD, NBS, EIA(P6) and others.

8.3.5 Fiber Optic Sensors

Fiber optic sensors, with their immunity to EMI and RFI, are finding wide use in industry process control as their additional advantages are appreciated, i.e., electrical isolation, insusceptibility to explosion, and remote location of electronics and probes. Furthermore, ease of installation and accuracy have expanded the applications from simple switches and counters to very precise analog sensors which can monitor the position, strain, temperature, pressure and a host of other properties.

Fiber optic sensors are now used in chemical analysis and power transmission, and eventually will be interfaced with fiber optic telecommunication systems to provide better capability for obtaining and handling process and automation data. There are four basic techniques for sensing: transmission, reflection, microbending and fluorescence. While all four can provide switching and analog function, transmissive sensors are generally used for the switches and counters, while the others find broader use in analog applications. All use disruption of a light path for switching and intensity variation for analog signals. Fiber optic intensity modulated equipment for industrial control is one benefit; the user can set up inexpensive *noncontact sensing* systems.[10]

A new form of an optical-fiber *data highway* telemetry system which has been developed for a number of industrial applications[11] can also be used in a utility system to enable a large number of separate data sources (sensors) to be input anywhere along the fiber cable, without involving optical connections (Fig. 8.8).[12] The technique of optical-fiber data-collection highway allows for multiple access points on one fiber, but does not involve breaking into the cable to input the data. The data-collection system operates using the fact that light within an optical fiber may be *phase modulated* using an *ultrasonic wave* passed across the fiber. Using appropriate electro-optic techniques, it is a straightforward matter to detect the consequent phase changes.

In the system of Fig. 8.8, ultrasonic frequency can be used to modulate the light from the laser. This frequency is different for each data-input point on the windmill turbine, and data is transmitted by means of *frequency-shift keying (FSK)* techniques. Thus, for example, 0.5 MHz may represent digit "1" and 0.505 MHz digit "0" for the channel associated with windmill turbine 1; 1 MHz and 1.05 MHz may represent "digital" values corresponding to the channel associated with windmill turbine 2, etc. Using sensor multiplexing techniques, the return "digital" signals arrive at the detector separated and are further decoded and processed in the control unit.

It is obvious that parameters in a group of windmill turbines, such as wind velocity, mechanical rotation, tension, electric current, phase, etc., can be continuously and accurately monitored without being affected by the high EMI environment, using fiber optic sensors in combination with an optical-fiber data collection highway system.

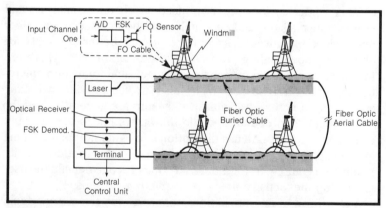

Figure 8.8—Use of Optical Sensors and Single Optical-Fiber Collection Line in a Windmill Energy System (Adapted from Ref. 12)

8.3.6 Fiber Optic Connectors Control EMI/RFI[13,14]

Precise alignment and mating of fibers are fundamental to the optimum performance of optical data links, distribution networks and long-haul transmission systems. Satisfying these demands requires reliable, demountable connectors with consistent and reproducible insertion-loss characteristics. Representative types of fiber optic connectors are shown in Fig. 8.9.

A variety of connectors and contacts are currently available to interconnect fiber optic subsystems, preventing EMI as regards the cable assembly. One of the most innovative connectors is a fiber optic contact which is a direct replacement for the contacts in the MIL-C-38999 Series I and II connectors. The contact gives immunity to EMI and RFI without requiring connector replacement or respecification. It fits into a size 16 cavity without modification, is moisture sealed to allow the use of electrical contacts in the same connector and features a precision jewel-ferrule alignment system which significantly reduces radial misalignment.

Another category is a series of environmental fiber optic connectors used in applications as varied as siesmic exploration and military field communications. This series, named FOMC, offers a design that permits disassembly and field cleaning while protecting against fungus, humidity and salt spray as well as a variety of corrosive elements such as hydraulic fluids. Aside from the hostile environment resistance, the immunity from EMI or RF emission ensures data security.

Optoelectronic Interfaces

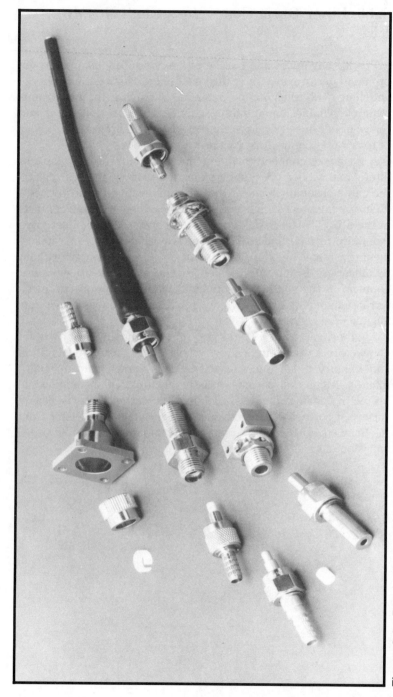

Figure 8.9—Representative Types of Fiber Optic Connectors *(Courtesy of Amphenol)*.

8.3.7 Radiation Effects on Fiber Optics

When exposed to a hostile, high-energy particle or radiation field, a fiber optic system may exhibit a half-life characteristic similar to that of atomic radiation decay. Radiation damages a material when its internal structure is changed by either ionization or bulk displacement. The energy required—called *penetration energy threshold*—depends upon the damage susceptibility total dose in *Rad-Si*.

The Rad-Si unit is a measure of the radiation energy absorbed in bulk silicon: 1 Rad-Si = 100 erg/gram = 0.01 joule/kilogram, or 1 Rad-Si = 3.0×10^7 electrons/cm^2 (for 1 MeV electrons). Considering published data [15,16] for various types of fiber optics and dose ranges, a curve for average optical transmission loss dB/m has been plotted as shown in Fig. 8.10.

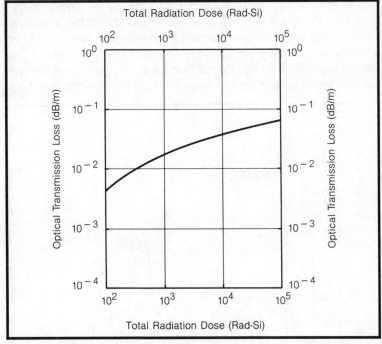

Figure 8.10—Average Optical Transmission Loss vs. Total Radiation Loss for Fiber Optics

8.4 Light Sources and Detectors in Fiber Optic Links

Sources and detectors used in fiber-optic transmission systems must be of sizes and configurations that are compatible with high-efficiency, low noise fiber optic cable and must be capable of signal modulation and detection at high rates.

8.4.1 Spectral Matching of Fiber Optics to Sources and Detectors

When choosing components such as emitting diodes, optical fibers and detector diodes, it is especially important to ensure that they are matched spectrally. The spectral maximum of the emitting diodes should agree with the attenuation minimum of the fiber and the maximum of spectral sensitivity of the receiving diode[17] (Fig. 8.11). This is particularly necessary when the user does not obtain complete fiber optic communications systems but prefers to build up an existing system from components of various manufacturers, requiring the components to be carefully matched.

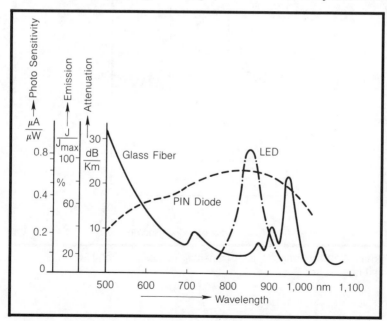

Figure 8.11—Spectral Compatibility of Emitters; Detectors and Low-Loss Fibers[17]

8.4.2 Light Sources and Their Interfaces

Semiconductor *injection lasers* and *light emitting diodes (LEDs)* based on the ternary material (GaAlAs) and quaternary material (GaIn-AsP) systems are used as optical sources in fiber-optic communication systems. The former emits in the wavelength range of 800 to 900 nm, while the latter emits at the range from slightly above 1,000 nm to approximately 1,700 nm. Exact emission wavelength depends on the device's material composition.[18]

The choice between the two types of optical sources will be determined by overall system requirements. In general, lasers are favored for wideband, long-haul, communication links because of their higher performance. LEDs are advantageous in short-haul, narrowband systems due to the relative simplicity of their drive and control circuitry. Table 8.2 presents a comparison between LEDs and diode lasers as applied to fiber optic systems.

Driving interface circuits for LEDs are simple. The series drive circuit of Fig. 8.12a, for example, minimizes the power supply current and is also recommended for driving multiple LEDs. However, the $V_{CE(SAT)}$ of the drive transistor must be taken into account when calculating R_1, the LED current determining resistor. On the other hand, the shunt drive circuit shown in Fig. 8.12b has the advantage of continuous power supply current drain and less power supply noise is generated. Shunt drive circuits can also be used with a lower power supply voltage and are recommended when driving single LEDs.

Table 8.2 LEDs vs. Lasers in Fiber Optic Systems

LEDs	Lasers
LEDs are characterised by:	In particular, lasers are characterised by:
• Longevity—10^6 hours • Low power consumption	• Directional emission • Outputs from milliwatts to as high as 10W
• Simplicity of construction • Relative cheapness	• High response speed (rise time)—typically 1 nanosecond
On the debit side, LEDs: • Deliver low power—in the microwatt to milliwatt range • Have slower response time—several nanoseconds at best Output power, launch and modulation efficiencies of some edge-emitting LEDs can approach those of a laser.	On the debit side, lasers are: • Expensive • Extremely temperature sensitive • Relatively short-lived. Although extrapolated lives may be 100 000 hours, most manufacturers do not provide guarantees at this level.

Optoelectronic Interfaces

The driving circuits of Fig. 8.12 are for digital transmission. It is also possible to use analog links. The analog transmitter differs from the digital in that the output of the LED diode at the transmitter must now be proportional to the input signal level to the transmitter interface. This arrangement is highly dependent upon the linearity of the LED, which normally gives a link performance figure of around 10 percent total harmonic distortion for about 80 percent modulation depth. Distortion and modulation depth can be traded, but range of transmission, signal-to-noise ratio performance or both will decrease with falling modulation depth. One method of reducing distortion lies with the use of optical feedback, where a small sample of the output is fed to a local receiver, the output of which is mixed with the system input in a differential amplifier.[19]

A laser transmitter is somewhat more complicated[1] than the LED transmitter, since the laser itself is a threshold device in which the threshold changes with temperature and aging. The laser driver dc-biases the laser slightly below threshold and adds enough signal current to bring the laser to an *on* state. Figure 8.13 shows a block diagram of a laser-driven circuit in which two factors (at the summing point) influence the modulation.[20]

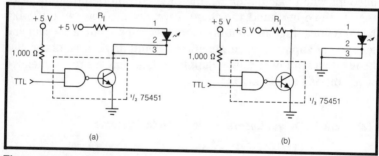

Figure 8.12—LED Interface Driving Circuits: (a) Series-Driven LED and (b) Shunt Driven LED

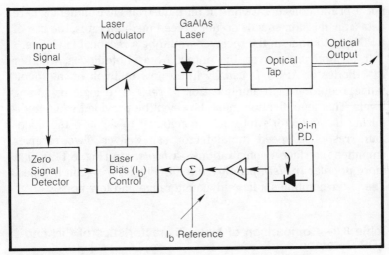

Figure 8.13—Block Diagram of Laser-Driven Circuit

8.4.3 Photodetectors and Their Interfaces

A photodetector "demodulates" an optical signal by generating a current proportional to the intensity of the optical radiation, thereby converting the variations in the optical intensity into an electrical signal. The most important characteristics of a photodetector are efficiency, speed, noise and physical compatibility. The most common detectors used in a fiber-optic system are the positive-intrinsic-negative *(PIN)* and avalanche photodiodes (APDs). The material used to construct these diodes reflects the wavelength range of interest. Silicon photodiodes can detect radiation from visible up to about 1 μm, germanium photodiodes respond up to 1.6 μm, GaAs photodiodes cover the range from 0.7 to 0.9 μm, while InGaAsP (or other III-V ternary and quaternary alloys) can respond up to 1.6 μm, depending on the material composition.[18]

The advantages of the PIN photodiode over an APD are low cost, circuit simplicity, wide bandwidth and low variation of sensitivity with temperature and bias. APDs have an inherent internal gain but require a high-voltage (up to 300 V) supply and temperature compensation circuitry. Table 8.3 shows some of the characteristics of PIN diodes and APD devices. Generally, silicon PIN diodes work best for short distance, low-cost systems, and silicon APDs are preferred when systems have bandwidth-distance products of up to 1,000 km-MHz and the detector's cost is

Optoelectronic Interfaces

not a prime consideration.[21] In an industrial environment where data transmission speeds do not exceed medium rates, the use of a PIN photodiode in the receiver would be a reasonable choice.[22]

In an optical receiver, the output from photodiodes, whether PIN diodes or APDs, is usually in the low millivolt or microvolt range, requiring high amplification to raise it to logic or analog levels. This amplification must, however, be provided while minimizing the detector's internal capacitance to achieve the maximum response speed the detector can deliver. Two receiver amplifier interface configurations—a *bootstrap* (Fig. 8.14a) or a more popular *transimpedance* circuit (Fig. 8.14b)—can be used. The latter combines a large dynamic range and low noise.

Table 8.3—Comparison of Main Characteristics of PIN and ADP Detectors

Parameter	p-i-n Photodiode	APD
Responsivity	0.5 A/W	75 A/W
Risetime*	1–10 ns	0.1–1 ns
Frequency Response	to 1 GHz	to 100 GHz
Internal Gain	1	50–500
Noise Equivalent Power	1×10^{-12} W/$\sqrt{\text{Hz}}$	1×10^{-14} W/$\sqrt{\text{Hz}}$
Reverse Bias Voltage	5–50 V	150–300 V
Lifetime**	10^7–10^8 hrs	10^6–10^7 hrs

*There are several types with risetime outside the indicated region
**Estimated

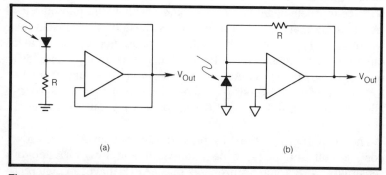

Figure 8.14—Optical Receiver Amplifier Circuitry: (a) Bootstrap Interface and (b) Transimpedance Interface

8.4.4 Snap-In Fiber Optic Links and EIA Interface Compatibility[23]

The *snap-in* fiber optic link, as it is called, combines a quick connection and disconnection scheme with fiber and optical device technology developed at Hewlett-Packard's Optoelectronics Division.

Standard communication level interfaces such as RS-232-C or RS-422 can easily be interfaced to the digital fiber optic link using standard interface integrated circuits. Figure 8.15 shows two snap-in fiber optic links: the EIA RS-232 and RS-422. Any handshaking signals required for communication exchange should follow the RS-232-C or RS-449 Specifications. Networks using multiple driven transmitters, or wire-OR receivers, are used in many networks for data interchange. Such a network, implemented with snap-in fiber optic links, is shown in Fig. 8.16.

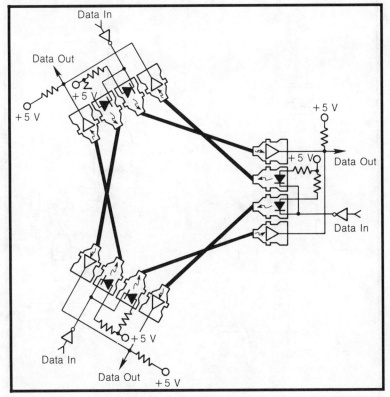

Figure 8.16—Multiple Series Driven Transmitters and Wired-OR Receivers in a Three-Node, Fully Connected Network (Courtesy of Hewlett-Packard Co.)

Optoelectronic Interfaces

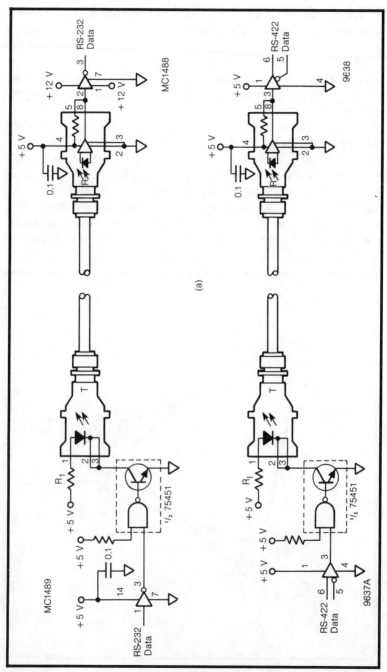

Figure 8.15—Data Interface with Snap-in Fiber Optic Links (a) RS-232 Data Interface and (b) RS-422 Data Interface (Courtesy of Hewlett-Packard Co.)

8.4.5 Flux and Bandwidth Budgeting[24,25]

In order to establish the *flux*- or *loss-budgeting* requirements of a fiber optic link, the characteristics of the receiver noise and bandwidth, coupling losses at connectors and transmission losses in the cable should be taken into consideration. For any of the previous fiber optic links, the flux which the transmitter must produce is determined from the expression:

$$10 \log \frac{\Phi_T}{\Phi_R} = a_O l + a_{TC} + a_{CR} + n a_{CC} + a_M \qquad (8.4)$$

where, ϕ_T = Flux (in μW) available from the transmitter
ϕ_R = Flux (in μW) required by the receiver
a_O = Fiber attenuation constant (dB/Km)
a_{TC} = Transmitter-to-fiber coupling loss (dB)
l = Fiber length (Km)
a_{CC} = Fiber-to-fiber loss for in-line connectors (dB)
n = Number of in-line connectors (excluding the end connectors)
a_{CR} = Fiber-to-receiver coupling loss (dB)
a_M = Margin (dB), chosen by the designer, by which the transmitter flux exceeds the system requirement.

Another technique helpful in selecting fiber optic system components is the *rise-time budgeting* or *bandwidth budgeting*. In using this technique, a system is partitioned into individual elements for bandwidth calculations, and then it is determined how each must perform to satisfy overall requirements. Because fiber-optic rise and fall times are roughly equal, data links can be characterized by rise time rather than bandwidth budgeting. These two parameters are related by the expression:

$$t_R = 0.35/\text{bandwidth}, \qquad (8.5)$$

with t_R in seconds and bandwidth in Hz. The receiver's output signal is given by:

$$t_{R(SYS)} = [t_R^2\text{(LED)} + t_R^2\text{(CABLE)} + t_R^2\text{(DET)}]^{1/2} \qquad (8.6)$$

8.5 Wireless Links Via Diffuse Infrared Radiation

Data transmission by direct or diffuse *infrared (IR)* radiation not confined in an optical cable is an interesting alternative to short-distance, medium-speed applications.[26-29] While preserving a high level of EMI immunity comparable to that of fiber optics, an IR link for confined space applications requires no cable installation, can be used as mobile links and offers substantial cost savings. However, a recurrent problem common to most IR systems operating in a closed area is the presence of interfering background events which may degrade the detector's detection capability.[30]

8.5.1 Infrared Link for Automated Factory Environment

A practical IR link for factory applications (Fig. 8.17) consists of a control station that initiates and transmits control and other function signals to a machine, robot or a cluster of automatic machines operating in the same area. It also accepts data by its receiving section that are transmitted from each machine. On the

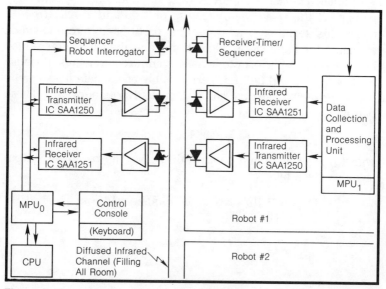

Figure 8.17—A Two-Way Infrared Communication System for Automated Factory Applications (Adapted from Ref. 29)

other hand, each machine is equipped with a processing unit as well as with receiving/transmitting hardware for two-way communication.[29]

In essence, Fig. 8.17 shows a no-wire communication multi-access channel where the transmission medium is diffusely scattered infrared radiation at 850 nm wavelength. Optical radiation transmitted from the LED (actually LED arrays) is diffusely scattered from the surrounding walls, ceiling and other objects in the room, thus filling the whole space with an optical signal carrier. Reception is nondirectional, i.e., the photodiodes (actually PIN arrays) receive the incoming radiation from a wide *field of view (FOV)*. This means that it is not necessary to have direct line of sight between transmitter and receiver. Therefore, the optical transmission is very insensitive to interruption and the system is very flexible.

The theoretical limitation of bandwidth is approximately 260 Mb × m/s, and it is due to multipath propagation and background noise produced by ambient light that reduces the transmission speed below 1 Mbps.[26] However, practical networks as tested in the laboratory have yielded much lower rates ($\approx$ 200 kbps) for a link operating up to 25 m. To achieve higher transmission of messages, since voice, data and video services may be needed, filtering, faster LEDs, PINs and probably more novel techniques will be required.

8.5.2 Sources of Interference and Other Detection Problems

In a system like the above, or in an instrument based on IR signal detection, receiver sensitivity limitations may be caused by background light, dark current, post-detection amplifier noise and transmitter imperfections. In a practical receiver operating at finite temperature, thermal effects can cause transients. The generated dark current can increase the required optical signal level for a desired receiver performance by several orders of magnitude above the quantum limit. Generally, in a well-designed receiver sensitivity is set up by input stage noise.

8.5.3 Background and Transient Light Effects

During an investigation regarding IR measurements in an office using special radiating fixtures and radiometric instruments[31] in order to establish the radiation levels for transmitting/receiving information for control and communication purposes, it was found that the real IR signal was largely affected by various background and extraneous light interference signals originating from thermal radiation, fluorescent and incandescent light as well as from the light of other rare gas lamps.

In an IR diffuse light channel, using LEDs as light sources (800 to 850 nm), the photodiodes at the receiver are exposed to the ambient light, introducing additional noise to the input circuit of the receiver amplifier. This is in contrast to fiber optic receivers. The most commonly encountered ambient light sources are daylight, tungsten and fluorescent lamps, to which a number of other lamps with special spectral outputs can be added, depending on the working area where an IR system will operate. Figure 8.18 shows representative curves for the relative spectral distribution of various kinds of light sources, along with the LED/PIN spectral characteristics. Knowledge of these spectra helps in the design of efficient filters.

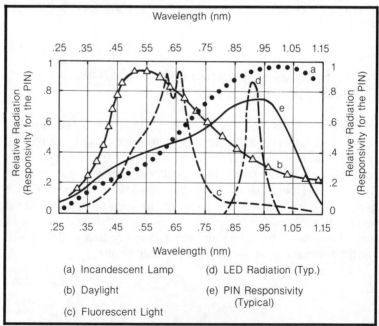

Figure 8.18—Relative Spectral Energy Distribution for Various Kinds of Sources and PIN Responsivity (Adapted from Ref. 30)

8.6 References

1. C. J. Georgopoulos, *Fiber Optics and Optical Isolators*, Don White Consultants, Inc., Virginia, 1982.
2. S. Waaben, *High Performance Optocoupler Circuits*, International Solid-State Circuits Conference Digest Technical Papers, Vol. XVIII, 1975.
3. P. Vettiger, *Linear Signal Transmission With Opticouplers*, IEEE Journal of Solid-State Circuits, Vol. SC-12, No. 3, June 1977.
4. M. H. El-Diwany, et al., *Piecewise Linear CAD Model for Avalanche Photo-detectors, Proc. IEEE*, Vol. 67, No. 8, August 1979.
5. H. Gass, *Providing Power for Gas Laser Tubes: Part 2, Photonics Spectra*, July 1984, p. 37.
6. J. Fagenbaum, *Optical Systems: A Review, IEEE Spectrum*, October 1979, p. 70.
7. A. Husain, *The Optical Interconnect: Microcircuit Problem Solver, Photonics Spectra*, August 1984, p. 57.
8. T. G. Giallorenzi, *Optical Communications Research and Technology: Fiber Optics, Proceedings of the IEEE*, Vol. 66, No. 7, July 1978, pp. 744–780.
9. M. C. Hudson and P. J. Dobson, *Fiberoptic Cable Technology, Microwave Journal*, July 1979, pp. 46–53.
10. D. A. Krohn and E. I. Vinarub, *Fiber Optics Invade Process Control, Photonics Spectra*, February 1984, p. 51.
11. B. Culshaw, et al., *Optical-Fibre Data Collection, IEEE Electronics and Power*, February 1981, pp. 148–150.
12. C. J. Georgopoulos, *Fiber Optic Sensors and Data Collection Highway Technique Applied to High Energy Systems*, EES '83—Energy and Environment Systems, Athens, Greece, August 29—September 2, 1983.
13. T. Ormond, *Fiber-Optic Connectors Cut Insertion Loss, EDN*, July 11, 1985, pp. 229–235.
14. L. Borsuk, *Fiber Optic Connectors Control EMI/RFI, RF Design*, July/August 1985, pp. 36–37.
15. G. H. Sigal, Jr., *Fiber Transmission Losses in High-Radiation Fields, Proceedings of the IEEE 68(10)*, (Tables I and III), October 1980, pp. 1236–1240.
16. A. N. Paolantonio, *Fiber Optic "Half-Life", Electro-Optical Systems Design*, August 1982, p. 33.

17. G. Knoblauch, *Fiber Optic Communications in Industry*, Siemens Components, Vol. XV, No. 3, 1980, pp. 144–150.
18. K. Y. Lau, *Semiconductor Sources and Detectors in Fiber Optic Systems*, Microwave Journal, April 1985, p. 97.
19. D. L. Jones, *An Engineering Approach to Fiber Optic Link Design*, Electronic Engineering, Mid-April 1980, p. 67.
20. E. E. Bash, H. A. Curnes and R. F. Kearns, *Calculate Performance Into Fiber-Optics Links*, Electronic Design, August 16, 1980, pp. 161–166.
21. J. Zuker, *Choose Detectors For Their Differences to Suit Different Fiber-Optic Systems*, Electronic Design, Vol. 9, April 26, 1980, pp. 165–169.
22. C. J. Georgopoulos and C. S. Koukourlis, *Fiber-Optic Link Design Considerations for Applications in Noisy Industrial Environments*, IEEE Transactions on Industrial Electronics, Vol. IE-31, No. 3, August 1984, pp. 209–215.
23. Hewlett-Packard, *Designing With the HFBR-0500 Series Snap-in Fiber Optic Link*, Application Note 1009, November 1980.
24. Hewlett-Packard, *Digital Data Transmission With Fiber Optic Systems*, Application Note 1000, Optoelectronics Designers Catalog, 1979.
25. J. Bliss and D. W. Stevenson, *Loss-Budgeting Techniques Simplify Fiber-Optic Links*, EDN, April 14, 1982, p. 159.
26. F. R. Gfeller and U. Bapst, *Wireless In-House Data Communication via Diffuse Infrared Radiation*, Proceedings of the IEEE, Vol. 67, No. 11, November 1979, pp. 1474–1486.
27. S. Donati, et al., *Optoelectronic Signal Transmission by Diffuse Radiations: Design and Performances*, Laser + Optik, November 2, 1981, pp. 70–72.
28. C. J. Georgopoulos, *The Use of Fiber Optics and Infrared Devices in Industrial Robots*, MECO '83—Measurement and Control, Athens, Greece, August 29–September 2, 1983.
29. C. J. Georgopoulos, *Controller-Infrared Network Links*, FOC/LAN '83, Atlantic City, New Jersey, October 10–14, 1983.
30. C. J. Georgopoulos, *Light Interference in IR Instrumentation and Sensing Systems Operating in Cinfined Spaces*, Intelligent Robots and Computer Vision—Cambridge Symposium on Optical and Electro-Optical Engineering, Cambridge, MA, November 4–8, 1984.
31. C. J. Georgopoulos and J. V. Korasis, *Modeling and Analyzing Small IR Antennas for Indoor Microbroadcasting Systems*, Proceedings of AMSE '84: Modeling and Simulation Conference, Athens, Greece, June 27–29, 1984.

General References

- W. Freeman, *TTL-to-RS-232 Adapter Needs No Separate Negative Voltage Supply*, Electronic Design, May 31, 1984, pp. 294–295.
- V. J. Maggioli, *The Impact of Fiber Optics on the Petroleum and Chemical Industry*, Industry Applications Society IEEE-IAS 1984 Annual Meeting, Chicago, IL, September 30—October 4, 1984.
- J. Javetski, *TI Quiets Interference by Switching Loads Using Fiber-Optic Interconnections*, Electronics, March 29, 1979, pp. 44–45.
- N. Albaugh, *Fiber Optic Modem Provides Good Noise Immunity*, Digital Design, September 1982, p. 26.
- E. Miller, *Introduction to Practical Fiber Optics*, IFOC, September 1980, pp. 63–71.
- Hewlett-Packard, *High Speed Fiber Optic Link Design With Discrete Components, Application Note 1022*, January 1985.
- T. Hiduka, et al., *Hollow Fibers for Infrared Transmission*, JEE, November 1983, pp. 72–77.
- M. M. Webb, *Fiber Optics Joins Advanced Microcontroller in Distributed Workstation Architecture*, Digital Design, August 1984, p. 112.
- C. J. Blickley, *Digital Process Measurements Transmitted by Fiber Optic Cables*, Control Engineering, January 1985, pp. 95–96.
- C. J. Georgopoulos, *Devices and Subsystems Connections in Modern Electronic Offices: Some Annoying Problems*, Melecon/85, Madrid, Spain, October 8–10, 1985.
- C. J. Georgopoulos, *Filtering Techniques for Free Channel IR Detection in Closed Working Areas*, Proceedings, SPIE's Cambridge Symposium on Optical and Optoelectronic Engineering, Cambridge, MA, October 26–31, 1986.

Chapter 9
Local Area Networks: Interfaces and Interconnections

As factories and offices move toward automation, *local area networks (LANs)* for sharing information and computer-related resources are bound to proliferate. A host of manufacturers have chosen varying topologies as the bases for their LANs, while designers differ in their preference for interconnection media and protocols. And where some systems are designed to use coaxial or twisted pair cables, others use fiber optics or wireless channels as their media.

9.1 Introduction

The topology of the particular LAN does not dictate which type of cable is actually used to link user stations. The majority of bus and ring structures use coaxial cable. Star networks exhibit a preference for existing telephone wire or shielded twisted pair. Regardless of topology, some manufacturers are commercially set up to deal with optical fibers, while others consider radio frequency (RF) or infrared (IR) for wireless end-drops in a LAN.

As in all electronic installations, unwanted electromagnetic energy from external transmitting sources can be conducted and radiated into the computing and other electronic elements of a LAN system, causing excitation of sufficient amplitude to disturb its operation. On the other hand, the LAN itself generates this noise, disturbing its own operation and that of other nearby equipment or systems. Techniques for reducing these effects in a typical LAN of data processing industry are discussed in Section 9.3. Another important topic, monitoring and alarm systems interface, is addressed in Section 9.4.

9.2 Definitions and Transmission Media for LANs

The primary objective of a LAN is to provide high-speed data transfer among a group of nodes consisting of data-processing terminals, controllers or computers within the area of a building, factory or campus environment. The various transmission techniques that are implemented within LANs can be generally categorized by the signaling scheme used to transfer the electrical energy onto the medium.

9.2.1 Some Fundamental Concepts

A fundamental understanding of the basic LAN configurations and control schemes that are prevalent today will help the reader to comprehend the more detailed discussions in this section and the subsequent sections. The first step is to provide some definitions for the current LAN concepts.

LAN (Local Area Network) can be defined as an information transport system for high-speed data transfer among a group of nodes consisting of office or industrial system terminals and peripherals, cluster controllers or computers via common interconnecting medium within the bounds of a single office building, building complex or a campus.[1-3]

Network Topology is a pattern of interconnection used among the various nodes of the network. There are two general types of topologies: (1) *Unconstrained* (Fig. 9.1) which is associated with a packet-switched network and (2) *Constrained* (Fig. 9.2), which is particularly suited to LANs. Such topologies are the bus, the star and the ring, which are defined below (see also Table 9.1).

The *bus* provides a bidirectional transmission facility to which all nodes are attached. Information signals propagate away from the originating node in both directions to the terminal bus. Each node is tapped into the bus and copies the message as it passes that point in the cable.[2,3] The use of repeaters at each endpoint

Fundamentals

can extend the bus network's range of operation. Network control is distributed. The best known bus LAN is *Ethernet*; another is its low-cost option *Cheapernet*.

A *star* network is implemented in point-to-point connection schemes which enable each node to exchange data with the central node. Most star LANs are offered by manufacturers of *private branch exchanges (PBXs)* or *computerized branch exchanges (CBXs)* where the central node acts as a high-speed switch to establish direct connections between pairs or attached nodes.

A *ring* is a closed-loop system with users connected to the cable or wire by nodes. In this case, an active repeater is usually placed between the node and the wire. Messages on a ring travel in one direction only. A ring LAN also offers high data rates over distances of some kilometers and is suitable for higher loading. Rings can be easily implemented with a variety of transmission media.

An *access method* is that part of a protocol that coordinates bandwidth use among all network subscribers. It ensures that only

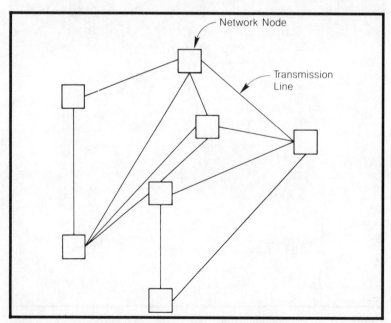

Figure 9.1—Unconstrained topology. In this configuration, each node receiving a message must make a routing decision.

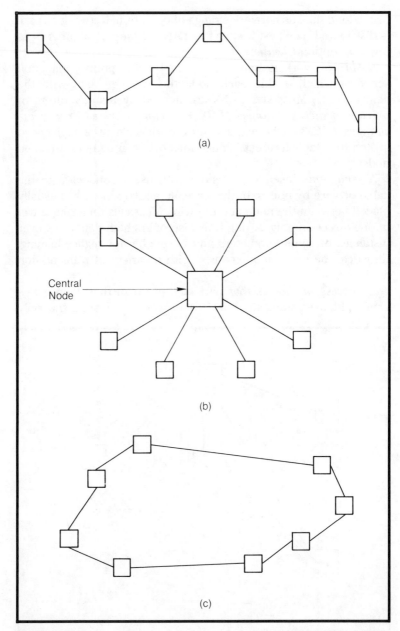

Figure 9.2—Representative Schemes of Constrained Topologies: (a) Bus (b) Star and (c) Ring

one station transmits at a given time, or if there is more than one, that proper recovery action is taken to provide correct data transmission. Two effective access methods are the Carrier Sense Multiple Access/Collision Detection (CSMA/CD) and the Token Passing.

The *Carrier Sense Multiple Access/Collision Detection (CSMA/CD)*[4] method can detect transmission collisions while the data are being transmitted. This minimizes bandwidth waste during collisions, but imposes a minimum size restriction on every frame to ensure that collisions are detected (Fig. 9.3a). A more serious disadvantage in collision detection is in its actual implementation. It must detect two simultaneous transmissions; a station's receiver must *listen* for others while its own transmitter is *talking*. Transceiver design is critical. Special cable and cable taps are often needed to minimize noise and impedance problems. Special installation and grounding practices which have been developed may require additional training of cable installers and modification of building codes. Several different systems using CSMA/CD are available commercially. The most notable is Ethernet, a joint project of DEC, Intel and Xerox.

Table 9.1—Comparison of Constrained Topologies
(Adapted from Ref. 4)

Topology	Advantages	Disadvantages
Bus	• Easily reconfigured • Media independent • Distributed control improves realibility	• Requires complex access control • Difficult to monitor network • Not suitable for fiber without active taps.
Star	• Simple protocol • Low incremental cost • Net information rate may be higher than transmission bandwidth • Easy network monitoring and control	• High initial cost • Reliability • Difficult to reconfigure (assuming cable or fiber)
Ring	• Simple protocol • Well understood • Ideal for fiber • Net information rate may be higher than transmission bandwidth	• Requires active taps • Reliability • Media dependent—not suitable for radio frequency or infrared.

LAN Interfaces

In *token passing*, a message in the form of a token is sent from one network node to another in one direction, where each node examines it. Each node has a specific time during which it can remove the token and strip or add a message to it. At this time, all other network nodes can only listen to the network. This process continues until the original sending node receives the token from the last network node and acknowledges that the intended recipient node got the message. Token passing thus guarantees access to each network node within a prescribed period of time, hence its highly deterministic nature. This is a critical quality for certain applications such as process control and other realtime tasks. The token-passing packet frame is shown in Fig. 9.3b. While both access methods are conceptually simple, there are several implementation challenges in the token scheme. These include network initialization, building and maintaining the logical ring and the resolution of fault recovery conditions. Centralized and fully distributed are two categories of solutions to these tasks. Table 9.2 shows a more detailed comparison between the CSMA/CD and token passing techniques.[5]

Transmission Techniques[2] are important because the digital information to be transmitted over a medium must first be electrically encoded so that the bits (1s and 0s) are distinguishable at the receiving node(s). The rate at which the encoded bit

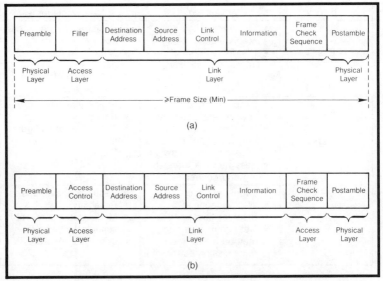

Figure 9.3—Packet Frames for (a) CSMA/CD and (b) Token Passing

Fundamentals

information is applied to the medium by a sending node is referred to as the transmission speed, expressed in bits per second. The encoded information is applied to the medium in one of two basic methods, commonly known as baseband or broadband signaling.

Baseband signaling is the simpler of the two methods; the encoded signal is applied directly to the medium as either a continuous stream of electrical pulses on a metallic medium or as a stream of light pulses on an optical medium. One node at a time may apply signals to the medium, resulting in a single channel over which signals from multiple nodes must be *time-multiplexed* to separate the energy. Baseband data rates exceeding 100 Mbps are possible. However, practical limitations, such as the rates at which the attached nodes can continuously send or receive information, and the maximum signal drive distance for a given data rate and medium results in LAN data rates substantially lower than 100 Mbps.

Unlike baseband, *broadband* transmission schemes employ analog signals and multiplexing techniques on the LAN medium to permit more than one node to transmit at a time. Multiple channels, or frequency bands, can be created by a technique known as *frequency division multiplexing (FDM)*. A typical broadband system has a bandwidth of 300 MHz which can be divided into multiple 6 MHz channels (as used in cable television signal distribution) with pairs of channels designated for bidirectional communications over a single cable. A standard 6 MHz channel can readily accommodate data rates up to 5 Mbps. Two adjacent 6 MHz channels can be used to provide a single 12 MHz

Table 9.2—CSMA/CD vs. Token Passing Comparison[5]

Function or Parameter	CSMA/CD	Token Passing
Cost (Small Systems)	1.0	0.75
Topological Flexibility	Linear Bus. Needs Active Tap. One Node per Tap.	Tree and Ring, Uses Passive Tap. Several Nodes per Tap
Medium	Baseband Coax	Media Independent
Priority Access	None	Supports Priority
Network Control	Probabilistic: No Real-Time Capability; Unstable under High Data Traffic Loads	Deterministic: Allows Real-Time Capability; Stable under All Load Conditions
Network Transactions	Speed/Distance Parameters Limit Max Throughput; Data Only, Digitized Voice Possible under Light Loads	No Speed/Distance Interdependence; Data and Digitized Voice under All Loads

LAN Interfaces

channel for data rates up to 10 Mbps. Broadband operation requires that a modulate/demodulate function be performed by radio-frequency modulators/demodulators (RF modes) at the sender and receiver respectively, resulting in higher cost per attachment than with baseband schemes. However, this enables broadband signals to be transmitted for longer distances. Table 9.3 provides a brief comparison of transmission modes characteristics.[6]

There are two major categories of *transmission media:* (1) Cable links and (2) wireless links, with the following subdivisions:

1. *Cable Links Using*:
   —Twisted pair cable
   —Coaxial cable, rigid, flexible
   —Fiber optic cable

2. *Wireless Links Using*:
   —Radio Frequency (RF) carrier
   —Infrared (IR) carrier

These alternative transmission media will be discussed in more detail in the next subsection.

Table 9.3—Transmission Modes Characteristics[6]

Transmission Mode	Distance	Noise Immunity	Passive Taps	Projected Interconnect Cost	Multiple Channels
Baseband	moderate	low	yes	low	no
Broadband	unlimited	moderate	no	moderate	yes

9.2.2 Implementation of Alternative Transmission Media

Criteria useful in comparing the relative performance of different transmission media include maximum useable distance, noise immunity, topology versatility, ease of installation and maintenance, and cost. In this section, representative types of transmission media are discussed and a comparison of their main characteristics is made.

Ethernet is one of the first *coaxial cable LANs*. The block diagram of a large Ethernet system is shown in Fig. 9.4. It consists of segments of 50 Ω coaxial cable. These segments can be up to 500 m in length, and interconnection of segments is provided by a repeater. A given network supports up to 1,024 stations. Each station transmits data in packets at a 10 Mbps data rate.

Simple *twisted pair connections* can support point-to-point communication in the 1 to 10 Mbps range over distances on the order of a kilometer between repeaters. As an example, the *ring wiring* scheme shown in Fig. 9.5 is a wiring concentrator which can be placed at strategic and protected locations in a building.

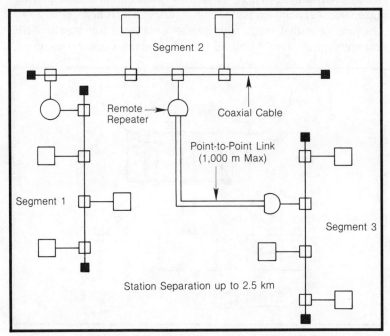

Figure 9.4—Large Ethernet System with Three Coaxial Segments

Offices are wired to concentrators in a radial fashion via so-called *lobes*. Here, shielded dual twisted-pair cables are used between wiring concentrator and stations, and shielded twisted-pair cables are used for the main ring with interconnect wiring concentrators.[7] In this particular case, fiber optic cables could also be used between some wiring concentrators, but wiring concentrators would need a power supply for the optical drivers.

Fiber optic cables are illustrated in Fig. 9.6 which shows a fiber-optic ring network. To achieve high reliability, the point-to-point fiber-optic links contain redundant paths. Configured as two separate data links, these paths permit ring operation despite a failure in the optical transmitter, receiver or cable. The redundant transmission links communicate in opposite directions around the ring. This arrangement permits continuous bus service between nodes, even in the case of a cut or open cable.[8] Of course, fiber optics can also implement other local network configurations. However, additional electrical and optical interface components may be required, and in each case cost and performance tradeoffs should be considered for best use of this technology.

Radio Frequency (RF) and microwave links are present in an office environment where it is now possible to use a private automatic branch exchange as a LAN controller. This *local network*, provided with appropriate interfaces for wireless RF communication, could link all office telephones and other termi-

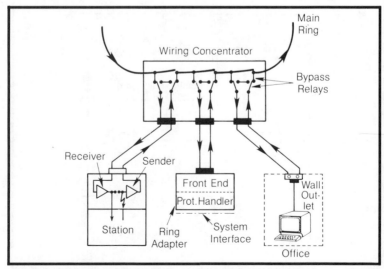

Figure 9.5—Wiring Concentrator and Station Interconnects in LAN Ring Configuration Where Shielded Twisted Pairs Are Used

nals to the switching equipment. A radio link has all the advantages of wireless communication schemes in terms of flexibility, hand-held control terminals and moving mechanisms in office, warehousing, manufacturing and instrumentation environments. Most commercial wireless telephones operate with radio carrier frequencies around 49 MHz using "frequency modulation." But to accommodate more users, the 862 to 960 MHz band allocated to mobile radio is more appropriate.[9,10] However, in order to over-

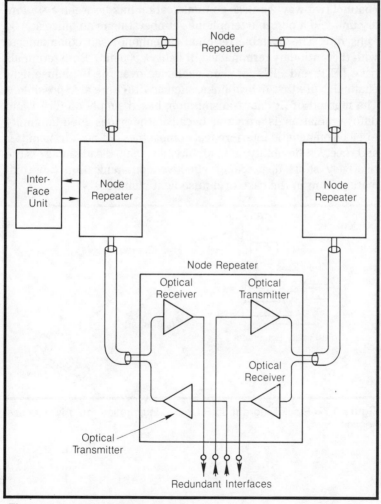

Figure 9.6—A Fiber Optic Ring System

LAN Interfaces

come the problem of multipath interference which is the result of reflections from the walls and objects within a building, spread-spectrum and other techniques must be used.[11] In addition, there are frequency leakage problems. For outside, relatively long distance communications, microwave links can be used. Microwave links can also be used in confined places (see Chapter 10).

Infrared links based on infrared diffuse light transmission can be used in a LAN effectively as wireless-drop configurations. Figure 9.7 shows a block diagram of an IR drop. This is a single-channel two-way wireless communication link between a stationary unit and a portable telephone set operating in an office at a 64 kbps rate. The portable terminal (telephone) can communicate with the stationary terminal via IR light (λ 900 nm). Both terminals have LEDs and PINs for data exchange over the IR diffuse light channel. Note that hybrid links, such as IR/IF, are also possible.[12] The method of IR data transmission, based simply on direct and diffuse radiation, is attractive because it provides good immunity to electromagnetic interference, compactness of both transmitter and receiver, flexibility and mobility. The main disadvantage is the relatively short distance of effective communication (approximately 50 m in the case of diffuse light channel).

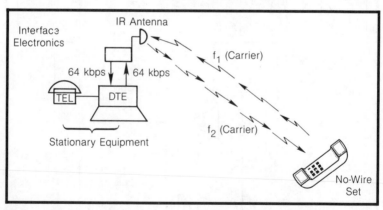

Figure 9.7—Block Diagram of IR Drop Configuration for LAN Wireless Applications

9.2.3 Comparison of Alternative LAN Transmission Media

Coaxial cable is one of the most common means of transmission, owing to its reasonable cost and high configurability. It allows transmission of a relatively high band. When broadband techniques are used, coaxial cable allows transmission of modulation-phase and PCM signals. It is also commonly used for channelled and simultaneous transmission in broadband radio frequencies.

The performance of a LAN based on twisted-pair cables is usually inferior to that of a baseband coaxial-based system, but often the cable is already in place in the form of the telephone network, and this will significantly reduce installation costs. In many applications, this is more valuable than the lack of high performance which may not be required anyway. The performance of the twisted cable improves if it is shielded or if the line has balanced drive.

Available fiber-optic components satisfy the needs of today's Ethernet-bus and token-passing ring networks.[8] Fiber optics are suitable for data bus distribution, owing to a high bandwidth capacity and a nonmetallic nature. The fiber optic bus does not take on the same architectural configuration as the coaxial bus because of the add/drop coupler elements and their associated connectors. Instead, a special class of couplers is needed to perform signal distribution in bus (multiterminal) data communications systems. There are two popular types of optical bus couplers, the *tee* and the *star* couplers.

It is important to note that optical fibers for the remaining 1980s will complement broadband and baseband coax installations, as shown in Fig. 9.8.[13] Regarding the wireless media, the RF/Microwave technique is subject to strict FCC regulations, and the IR technique is not fully proven yet. Table 9.4 compares the most important characteristics of LAN media. Table 9.4 comes from Ref. 14, but it was modified to include infrared (radiation) of diffused light for confined spaces.

Of course, not all LAN systems are confined to any one of the above media. For example, hardwire techniques for wideband local transmission capacity may include copper wires or fiber-optic cables or a combination of these (Fig. 9.9) with optical couplers and electro-optical transducers. A particular case is

LAN Interfaces

Datapoint's *Arc Net* system that currently exists in over 8,000 locations worldwide. This LAN supports a number of transmission media including coaxial cable, three-pair twisted wire and fiber optics.

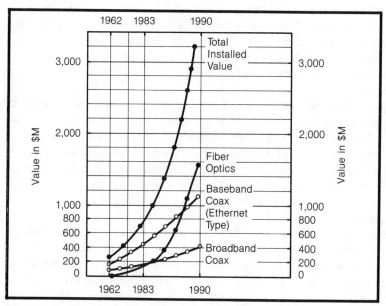

Figure 9.8—Value of the Installed Base of Local Nodes by Medium. 1982-1990 (Source: Internation Resource Development[13])

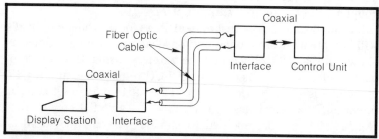

Figure 9.9—Fiber Optic/Coax Digital Hybrid Link

Media Comparison

Table 9.4—Comparison of LAN Alternative Transmission Media*

Medium	Range	Bandwidth	Loss (20 MHz)	EMI/RFI Immunity	Security	Cost	Comments
Twisted pair (shielded)	< 1200 km	<100 kHz	High	Very good	Fair	Low	Inexpensive, limited
Twinax	<1 km	<200 Mhz	50 to 100 dB/km	Very good	Fair	High	"Controlled impedance," can be 802-compatible.
Multiconductor flat cable	<100 m	<10 MHz	High	Poor	Fair	High	Shielding available, Can be byte parralel
Coaxial cable	<2 km	<400 MHz	6.5 dB/km 60 dB/km	Fair	Fair	Very High	Can be 802-compatible
Fiber optics	2 to 10 km	>500 MHz-km	1 to 8 dB/km	Excellent	Very good	$1 to 5 million	Cost continue to decline
Telephone line (PABX, PBX)	—	<60 kHz	High	Poor	Poor	Low	Low cost, installed base
Infrared (radiation) Line-of-Sight	300 m	<20 MHz	High	Good	Fair	Currently high, Potentially Low	(Free space) line of sight, reduced range in fog) moisture dependent)
Infrared (radiation) Diffuse[1]	<50 m	<1 MHz	High	Good	Good	Medium, Potential- by low	Susceptible to background light noise
Microwave (radiation)	<50 km (horizon)	>100 MHz	Inversely proportional to distance	Poor	Fair	High	(Free space) line of sight, subject to scattering, link loss depends on antenna size

*From Ref. 14 with the addition of diffuse infrared radiation (1).

9.2.4 Standardization: The ISO Open Interconnection Model

The Institute of Electrical and Electronic Engineers, the American National Standards Institute, the International Federal of Information Processing Societies, the Electrotechnical International Committee and the Electronic Industries Association have set up committees to work out standards for local data communications networks. The committees have a model of the architecture developed by the International Standards Organization (ISO). It is the Open Systems Interconnection (OSI) model, shown in Fig. 9.10, developed to make the standardization of data communication manageable.

The OSI model is a concept for developing compatible communications between heterogeneous devices. The model consists of seven layers which modularized the different functions and services required. Actually, the model identifies the hardware and software needed for two devices to communicate with each other. Its seven layers are: application, presentation, session, transport, network, data link and physical. Equipment can be designed for different purposes, but only equipment with the same complexity—or the same number of OSI levels—will be able to communicate.

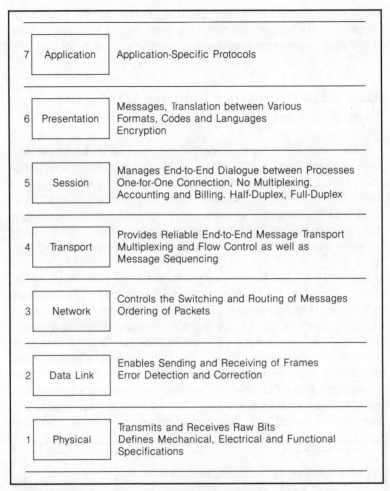

Figure 9.10—ISO Open Systems Interconnection Model

9.3 Controlling Interference in LAN Systems

As in all electronic installations, unwanted electromagnetic energy from external transmitting sources can be conducted and radiated into computing and other electronic equipment of a LAN system, causing excitation of sufficient amplitude to disturb its operation. On the other hand, the LAN itself generates this noise

LAN Interfaces

and disturbs its own operation and that of other nearby equipment or systems. Techniques for reducing these effects in a typical LAN of the data processing industry are discussed in this section.

9.3.1 Data Processing Industry LANs: The "Antenna Farm"

There have been three principal approaches to LAN systems, each evolving from a different consideration and historical standpoint (Fig. 9.11). For the data processing industry, it can be said that it has designed the actual devices that are interconnected in local area network applications. Experience in interconnection in this industry has evolved from backplane buses and parallel lines between nearby devices to serial data links connecting computers.[15]

The approach has focused on the most efficient way of interconnection based on the performance and hardware requirements of the devices themselves. In such a system, however, the application of square wave or other pulse signals and high switching frequencies leads to radiations of electromagnetic fields containing frequency components up to the UHF region. Consider, for example, a video display unit of a terminal device, where its signal is amplified from TTL levels to several hundred volts before it is fed into the cathode ray tube (CRT). The radiation originating from this terminal will join other radiations coming from the

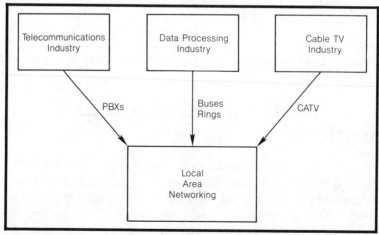

Figure 9.11—Major Types of Local Area Networking[15]

power supply, the CPU-peripheral interconnecting cables and other equipment, thus forming an *antenna farm*[16] in a digital system installation (Fig. 9.12), which can be part of a LAN system.

Failure to follow proper noise suppression and grounding rules in the design, installation and interconnection of digital equipment is increasing problems in the field. Hence, the work for taming EMI in an antenna farm of a LAN system has to be methodic and efficient. The methodology suggested here includes the treatment of EMI problems by focusing on power supply line-to-load conditioning, interface devices and CPU-terminal/peripheral interconnection lines, noise immunity improvement by correct choice of protocols and installed cables vs. environment.

Of course, the best solution is to use fiber-optic cables and optical isolators whenever possible. This subject will not be discussed further here, since it has been covered in Chapter 8.

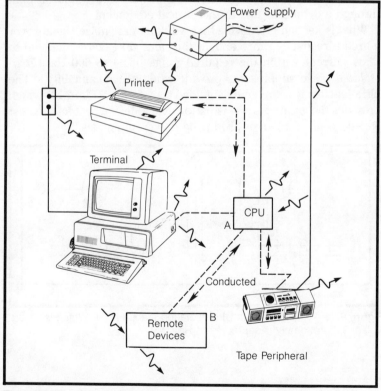

Figure 9.12—EMI Contributions of Power Supply, Digital Equipment and Interfacing Cables to "Antenna Farm" in a LAN System (Adapted from Ref. 16)

9.3.2 Power Supply Line-to-Load Conditioning

Most computers have incorporated dc power supplies with large input bulk dc filters. These filters, however, are ineffective against the majority of noise transients; therefore, filters are often inserted externally between the ac power source and the critical load. But these filters have serious disadvantages too, primarily in that they are frequency discriminating.

Another technique used for reducing noise-related computer problems is to provide the computer with a dedicated power line, i.e., one which is not shared by any other electrical equipment. However, a dedicated power line will protect a computer against noise generated by electrical devices located within the computer facility, but it provides no protection against noise transients that originate outside the facility. Such signals travel along the main distribution line, down the building power feeder and straight through a dedicated line to the critical equipment.

What is needed is a device that will attenuate noise signals over a broad frequency range and, at the same time, allow the load to draw current within the required range of 50 to 600 Hz. *Ultra-isolated* noise suppressors provide exactly this capability.[17] The ultra-isolator is a triple-box shielded transformer (Fig. 9.13) which provides the greatest noise isolation available and protects against 88.5 percent of all error-producing power anomalies.[18]

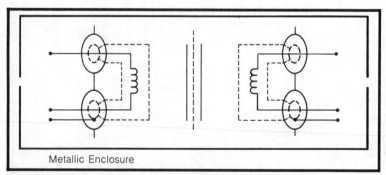

Figure 9.13—Shielding in an Ultra-Isolation Transformer, Which Isolates Noise Bidirectionally (Adapted from Ref. 18)

9.3.3 Interface Devices and CPU Terminal/ Peripheral Interconnection Lines

Figure 4.5 (in Chapter 4) shows an interconnection scheme between the central processing unit (CPU) and a remote CRT terminal along with other intermediate device connections including a repeater. This scheme could be imagined as replacing the section A—B in Fig. 9.13. It is a typical *bus* application using a combination of drivers, receivers and transceivers according to EIA RS-485 standard. This configuration is good for distances up to 1,200 m and 100 kbps data transfer rates. In the case of small computer systems, the Small Computer System Interface (SCSI) can be used. It is capable of interconnecting small computers and intelligent peripherals such as rigid disks, flexible disks, magnetic tape devices and printers. This interface need only operate over 15 m and 32 Mbps data rates. Of course, there exists a host of other bus systems with their associated interfaces, such as the Standard (STD) Bus, the General-Purpose Interface Bus (GPIB), the Multibus and the VME Bus. Their selection depends on their particular functions (address buses, data buses and control buses), their architecture and their cost (see also Chapter 5).

In high-performance systems, it is important that the cables have a controlled impedance. Choices among the various controlled impedance cables include coaxial twisted pair, standard and special ribbon, and flexible flat transmission. The choice of cable, as well as the arrangement of signals and grounds in the cable, will determine transmission path characteristic impedance. Low-cost approaches such as standard twisted pair and standard unshielded ribbon do not permit a characteristic impedance below about 80 Ω; higher-cost, special cable approaches allow well-controlled characteristic impedance ranging from 50 to 150 Ω. The various problems associated with interconnecting cables, including ground loops and cable shielding have been discussed in Chapter 3.

9.3.4 Noise Immunity Improvement by Correct Choice of Protocols[3,19,20]

Choosing the right standard requires the consideration of a number of factors, including the distance of data transmission, the data rate and volume and the amount of noise that might be added

LAN Interfaces

to the signal before it reaches its destination. Of course, high priority is given to the protection of the system against lightning, high levels of RFI and EMI. As discussed in Chapter 3, EIA has developed several specifications to standardize the interface in data communication systems. Many of these specifications apply to LAN systems too, especially for the terminal and peripheral connections. Tables 9.5 and 9.6 show available line drivers and line receivers, respectively, which meet standard protocols.

One way to guard against the effects of noise is to use *hysteresis* in a receiver. To accomplish this, the positive signal must exceed a value of V_{T+} volts before it is recognized as a high, and the negative-going signal must fall below a value of V_{T-} volts to be recognized as a low. The difference between the V_{T+} and V_{T-} is the amount of hysteresis. As an example, the NE5180 octal line receiver from Signetics has built-in hysteresis.

Another method is to use frequency shift keying (FSK) in modulation. Since the FSK transmitter drives the transmission line with a single-frequency sine wave, there are no harmonics to be attenuated and phase-shifted. The received signal is a sine wave. While a baseband signal reduced to a sine wave is undesirable, the sine wave is desirable in FSK transmission. The distributed constants of the transmission media tend to make all signals

Table 9.5—Representative Line Drivers that Meet EIA Standards

Device	Configuration	Pins	Manufacturer	Protocols Supported
MC1488	quad	14	Fairchild, Hitachi,, Signetics, National Semiconductor	RS-232C
AM26LS29	quad	16	Advanced Micro Devices, Signetics	RS-423A
NE5170	octal	28	Signetics	RS-232C RS-423A
MC3487	quad	16	National Semiconductor, Motorola, Texas Instruments	RS-422A RS-423A
µA9636A	dual	8	Texas Instruments	RS-432A
µA9616	triple	16	Fairchild	RS-232C
SN75150	dual	8	Fairchild, National Semiconductor,, Texas Instruments	RS-232C RS-422A

approach a sine wave, eliminating signal distortion. The IEEE 802.4 standard calls for this type of FSK system. FSK systems, built with devices such as the Signetics NE5080/5081 FSK transmitter/receiver pair, are designed to work with the 802.4 IEEE standard for a *token passing bus*, and they find applications with twisted-pair lines and fiber optics. IEEE 802.4 calls for a maximum data rate of 5 Mbps and a maximum transmission distance of 25,000 ft. For the FSK case, this standard calls for a 2 Mbps transmission rate and an error rate of 10^{-9}.

The IEEE 802.3 (*Ethernet*) standard specifies a maximum transmission distance of 10,000 ft. A low-cost variant has been added to the IEEE 802.3 document, called *Cheapernet*. The DP8390 is one chip set that forms a complete IEEE 802.3 node (and therefore a Cheapernet node). It provides reduction in circuit complexity, board area and power requirement. By appropriate partitioning of the functions required and choice of process technology, the three-chip set contains all the electronics for an IEEE 802.3 node.

The DP8392 *Coax Transceiver Interface (CTI)* implements the driver, the receiver and the collision detect circuitry and is in direct contact with the coax cable. The DP8391 *Serial Network Interface (SNI)* carries out the Manchester encode and decode of the serial data stream and interfaces to the transceiver cable. The DP8390 *Network Interface Controller (NIC)* executes the commu-

Table 9.6—Representative Line Receivers that Meet EIA Standards

Device	Configuration	Pins	Manufacturer	Protocols Supported
MC1489/A	quad	14	Fairchild, Signetics, National Semiconductor Texas Instruments, Hitachi, Motorola	RS-232C
SN75154	quad	16	Fairchild, Texas Instruments	RS-232C RS-422A
NE5180	octal	28	National Semiconductor Texas Instruments	RS-422A RS-423A
MC3486	quad	16	National Semiconductor, Texas Instruments	RS-422A RS-432A
SN75173B	dual	8	Texas Instruments	RS-422A
µA9637A	dual	8	Fairchild, Texas Instruments	RS-422A RS-423A
AM26LS32	quad	16	Advanced Micro Devices, National Semiconductor	RS-422A RS-423A

LAN Interfaces

nications protocol and provides the host system interface. It has built in buffer management hardware. The coaxial transceiver is the most difficult to integrate onto a single chip. The reason is that it combines high-precision analog functions with high-speed ECL-based digital circuitry. This provides all the functions required by an 802.3 *Medium Access Unit (MAU)* except signal and power isolation. It contains a driver, a jabber detector, a receiver and collision detector and a "heartbeat" generator (Fig. 9.14).

Dc isolation of power and signal connections between the transceiver and data terminal equipment (DTE) requires some extra components. For add-on designs, a dc-to-dc converter can be used and the CTI's low current drain will help keep costs down. Signal isolation is provided with pulse transformers. In Cheapernet designs, these are directly connected on the other side to the second chip in the set, the DP8391, as shown in Fig. 9.15. In Ethernet, the transceiver cable lies between. In either case the connection does not require additional isolation at the DTE end. The serial network chip's differential drivers and receivers meet all of the connection's short circuit, breakdown and common-mode requirements.

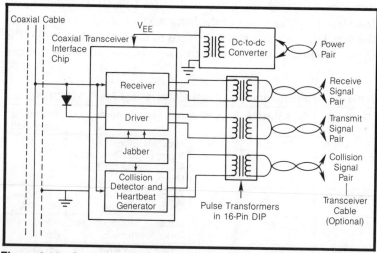

Figure 9.14—Coaxial Transceiver Interface Chip (DP8392) Isolation with Transformer. *(After Ref. 20)*.

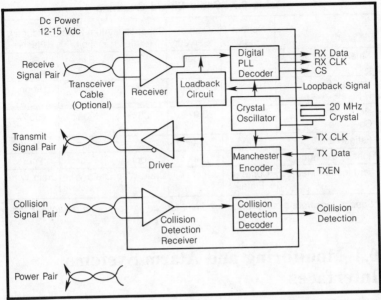

Figure 9.15—Signal Isolation in Serial Network Interface Chip (DP8391) by Differential Line Drivers—Line Received (Adapted from Ref. 20)

9.3.5 Installed System Cables vs. Environment

As cables are improved, the demands on them increase to meet various environmental conditions while maintaining high *reliability* at low cost. Factors affecting reliable performance of installed cables include the *aging* effects of moisture, light, low and high temperatures, corrosion, solar radiation and stress cracking.

Cable, during handling and installation and while in service, is often exposed to rigorous mechanical abuse from cable handlers, equipment, soil, animals and in some cases, insects. Depending upon the degree of protection required, a range of sheath designs are available for the system designer. Of course, the *qualities* of a *cable shield* play an important role in the cable's performance and reliability, too. Table 9.7 shows the qualities of cable shield for various types of materials, based on an early EIA investigation.

LAN Interfaces

Table 9.7—Cable Shield Qualities

	Copper Braid	Copper Served	Conductive Textile	Alum-Mylar Foil	Conductive Plastic
Shield Effectiveness A-F	Good	Good	Fair	Exc.	Fair
Shield Effectiveness R-F	Good	Poor	Poor	Exc.	Poor
Normal % of Coverage	60-95%	90-97%	100%	100%	100%
Fatigue Life	Good	Fair	Exc.	Fair*	Good
Tensile Strength	Exc.	Fair	Poor	Poor	Poor
Termination Method	Comb & Pigtail	Pigtail	Drain Wire	Drain Wire	Drain Wire

*Special techniques can provide excellent fatigue life for use in retractile cords.

9.4 Monitoring and Alarm Systems Interfaces

In modern network systems, the more routine functions like monitoring, fault isolation or bypassing and alarm management, must be accomplished without the need for operation attention.[21] This means that proper interfaces must be used which will help materialize the above functions without impairing the performance of the systems.

9.4.1 Automatic Monitoring and Restoration

The interface to a digital link in a LAN is via *Data Terminal Equipment (DTE)* or *Data Circuit-Terminating Equipment (DCE)*, usually interpreted as a *modem* or a front-end processor. Monitoring the performance of these interfaces must be the system designer's first consideration. The earliest and most basic approach to managing the network starts with equipment which resides at the central site. The network manager must first provide access to both the digital and analog interfaces of all modems in the network. Modems are in the unique position of bridging the analog and digital domains at every network site. In addition to the normal main data channel supported by the modems, they create a special test channel in both domains. An important feature of the network-control system is the test module which resides at every terminal interface and communicates over the test channel. This test module must recognize its own address when instructed, then accept commands and perform required functions.

Figure 9.16 shows an office network system with two local groups of data lines which report *Network Interface Unit (NIU)* circuit monitoring and alarming conditions back to a control network in the workstation. If a data monitor is used which has automonitoring capabilities, no hands-on readjusting of the data monitor is required, regardless of what type of traffic protocol is monitored, even synchronous. Automonitors can automatically detect the protocol, speed, bits per character and traffic on the port.

Switches are available which have monitoring capability for transmitting data, receiving data and *Carrier Detect (CD)*. It is even possible to get standard A/S switching with an adjustable audible alarm on RS-232-C leads, such as the CD. Thus, if the CD goes down (indicating that either a terminal line is lost or a modem has problems), the operator would be immediately notified. In general, an ideal system in an office would automatically scan and monitor for present alarm conditions, run autodiagnostic tests to isolate a problem and, once the source of the problem is located, perform rerouting or equipment substitution via intelligent switching to restore communications within the network.

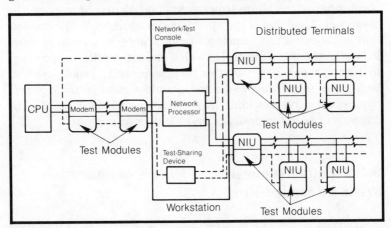

Figure 9.16—Simple Office Network with Individual Test Modules. Each test module recognizes its own address when instructed, then accepts commands to perform its assigned functions.

9.4.2 Alarm Management in Distributed Control Systems

In the case of process control systems, many of the alarm conditions generated during a process disturbance may be interrelated. Nuisance-reporting may therefore result from any one system going into alarm which is generated by digital inputs that cycle in and out of alarm and nuisance alarms that are generated by inoperative process equipment.[22]

Although many process control systems have the *built-in functionality* to provide *alarm management*, implementation is not always obvious. In fact, alarm management should be thought of as a function that would allow an implicit recognition of alarm patterns. Implementation schemes of alarm management techniques, in terms of operating a plant, deal with certain main objectives:[22] elimination of alarms for nonoperating plant equipment, elimination of alarms generated during shutdown of a plant or portions of a plant, elimination of all but the most important alarms in cases where one alarm is the direct cause of several others and segregation of alarms into priorities. This ensures that the most important alarms will receive first attention.

In general, the above objectives can be accomplished through alarm prioritizing, alarm suppression and alarm reduction at the instrument level and at the application module level. The application module is a programmable point processor used to perform higher level control strategies which are not possible or practical in the controller. To ensure that no impairing EMI problems occur because of monitoring and alarm circuitry, good grounding techniques and cable shielding (see Chap. 3) must be a continuous concern of both the designer and the user.

9.4.3 Fault Bypassing in Ring Configuration With Alternate Ring

Any reliable system must be decentralized so that its operation does not depend upon performance or availability of unique system elements such as master clocks or token management hardware. Token ring systems with fully decentralized clock distribution, plus token management and error recovery, can and are being designed.

Fault Bypassing

Advanced *wire center* concepts make rapid fault isolation and bypassing possible. The wire centers play the same role in the communication network as a circuit breaker panel plays in a power distribution network. If a defective node tries to attach itself to the ring, the wire center causes that node to be bypassed without disrupting the whole ring.[23]

A separate technique is necessary if reconfiguration is required as a result of faults occurring within the ring segments interconnecting *wiring concentrators*. Alternate backup links can be installed between the wiring concentrators in parallel with the peripheral links (Fig. 9.17).[2] If a fault occurs in the ring segment between two wiring concentrators, wrapping the principal ring to the alternate ring within the two wiring concentrators restores the physical path of the ring. The figure shows four wiring concentrators as they would be configured to bypass such a fault with both a principal and an alternate ring. The signals on the alternate ring are propagated in the direction opposite to those on the principal ring, thus maintaining the logical order of the nodes.

LAN Interfaces

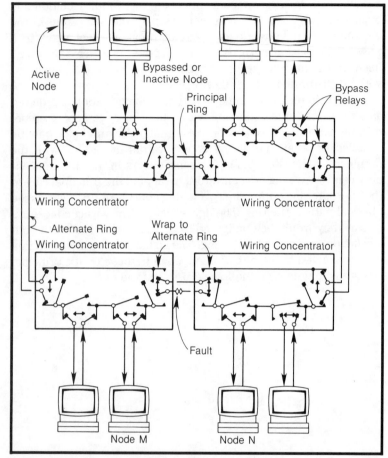

Figure 9.17—Ring Configuration with Alternate Ring. A fault on the ring between two wiring concentrators can be bypassed by wrapping the connections within the wiring concentrators, thereby using the alternate path and maintaining the same node ordering that existed before the fault occurred. (Reproduced from Ref. 2 by Permission, Copyright 1983 by IBM Corp.)

9.5 References

1. R. C. Dixon, N. C. Strole and J. D. Markov, *A Token-Ring Network for Local Data Communications, IBM Syst. J. 22*, 1983, pp. 47–62.
2. N. C. Strole, *A Local Communications Network Based on Interconnected Token-Access Rings: A Tutorial, IBM J. Res. Develop.*, Vol. 27, No. 5, September 1983, pp. 481–496.
3. D. C. Clark, et al., *An Introduction to Local Area Networks, Proceedings of IEEE*, Vol. 66, No. 11, November 1978, pp. 1497–1517.
4. M. Stieglitz, *Local Network Access Tradeoffs, Computer Design*, October 1981, p. 163.
5. G. Kotelly, *Local-Area Networks: Part 1—Technology, EDN*, February 17, 1982, pp. 109–117.
6. E. Pevovar and B. McGann, *Sorting Through the LAN Morass, Digital Design*, November 1982, p. 54.
7. W. Bux, et al., *Architecture and Design Considerations for a Reliable Token-Ring Network*, 4th World Telecommunication Forum, Geneva, Switzerland, October 29—November 1, 1983.
8. J. R. Jones, *Consider Fiber Optics for Local-Network Designs, EDN*, March 3, 1983, pp. 105–107.
9. H. H. Hoffman and D. C. Cox, *Attenuation of 900 MHz Radio Wave Propagation Into a Metal Building, IEEE Transactions on Antennas and Propagation*, July 1982, pp. 808–811.
10. S. E. Alexander, *Radio Propagation Within Buildings at 900 MHz, Proceedings of ICAP/83*, pp. 177–180.
11. C. J. Georgopoulos, *Spread-Spectrum Techniques for Reliable and Private Communications in the Office of the Future*, Proceedings of the DIGITECH '84 IMACS European Meeting, Patras, Greece, July 9–12, 1984.
12. C. J. Georgopoulos and V. C. Georgopoulos, *IR/RF Transmission System With Common Modulating Circuit, Electronics Letters 23*, November 8, 1984.
13. International Resource Developments, *Value of the Installed Base of Local Nodes by Medium, 1982–1990, Fiber Optics and Communications Newsletter*, August 1982, p. 8.
14. N. Makhoff, *Demand Grows for Integrated Open Networks, Computer Design*, September 1, 1985, pp. 51–61.
15. S. Joshi, *Understanding the Standardization Process for Networks, Communications Engineering International*, March 1984, p. 29.

16. C. J. Georgopoulos, *EMI Control in the Installation and Interconnection of Digital Equipment*, EMC Technology, March/April 1986, pp. 55–63.
17. G. Dang, *Ultra-Isolation Devices Yield Better Protection from Electrical Noise*, Computer Technology Review, Spring 1985, p. 187.
18. P. K. Halliman, *Power Conditioners Cut System Costs*, Digital Design, January 1982, pp. 68–71.
19. N. C. Gray, *Correct Choice of Protocols Increases Noise Immunity*, Computer Design, November 1, 1985, p. 125.
20. S. Britnell, *LSI Chip Set for Cheaper Networks*, Communications Systems Worldwide, November/December 1984, pp. 63–65.
21. C. J. Georgopoulos, *The Evolution of Automatic Monitoring in the Office of the Future*, Proceedings of First European Workshop on the Real Time Control of Large Scale Systems, Patras, Greece, July 9–14, 1984.
22. P. Schellekens, *Alarm Management in Distributed Control Systems*, Control Engineering, December 1984, pp. 60–64.
23. H. C. Salwen, *In Praise of Ring Architecture for Local Area Networks*, Computer Design, March 1983, p. 183.

General References

- J. Williamson, *Local Area Networks: A Review*, Communications Engineering International, June 1982, p. 34.
- M. K. Webb, *Fiber Optics Joins Advanced Microcontroller in Distributed Workstation Architecture*, Digital Design, August 1984, p. 112.
- S. P. Joshi, *Making the LAN Connection With a Fiber Optics Standard*, Computer Design, September 1, 1985, pp. 64–69.
- J. E. Deavenport, *EMI Susceptibility Testing of Computer Systems*, Computer Design, March 1980, pp. 145–149.
- R. H. Cushman, *Hands-On Network-Design Project Gives Insight Into LAN Features*, EDN, March 22, 1984, p. 219.
- G. A. Katopis, *Delta-I Noise Specification for a High Performance Computing Machine*, Proceedings of the IEEE, Vol. 73, No. 9, September 1985, pp. 1405–1415.
- J. L. Steenburgh, *The Incremental Method of Computer Shielding*, Computer Design, October 1969, pp. 64–69.
- D. R. J. White, K. Atkinson and J. D. M. Osburn, *Taming EMI in Microprocessor Systems*, IEEE Spectrum, December 1985, pp. 31–37.

- G. Dash, *Minimizing EMI at Minimal Cost in Computing Equipment, Electronics*, August 25, 1983, pp. 131–134.
- E. B. Rechsteiner, *Clean, Stable Power for Computers, EMC Technology*, July-September, 1985, p. 67.
- D. T. Y. Wong, *Controlling Electromagnetic Interference Generated by a Computer System, Hewlett-Packard Journal*, September 1979, pp. 17–19.
- W. H. Lewis, *Effective Computer Installations Require System Noise Protection, Computer Technology Review*, Fall 1984, p. 187.
- P. Zahra and C. M. Kendall, *EMC Control in Main Frame Computing Systems*, IEEE 1985 International Symposium on Electromagnetic Compatibility, pp. 432–436, Wakefield, MA, August 20–22, 1985.

Chapter 10
RF and Microwave Systems Interfaces

RF and microwave techniques are widely used in therapeutic procedures, radars, electronic warfare and countermeasures, guidance systems and satellite communications. Other applications include microwave cookers, speed detectors, railway traffic control and intruder alarm systems. In designing these systems and their interfaces, special care must be taken to reduce electromagnetic interference, which is the hidden agent in every RF/microwave design.

10.1 Introduction

Radio Frequency (RF) is a frequency at which coherent electromagnetic radiation of energy is useful for communication purposes. The spectrum of radio frequencies extends from about 10 kHz to about 300 GHz. The portion of this spectrum that covers the range from about 1 GHz to 100 GHz corresponds to *microwave* frequencies.

Section 2 of this chapter deals with *friendly fields* of RF and microwaves as they apply to medical electronics. Problems that arise because of mismatched interfaces are discussed in Section 3, where various matching and balun circuits are presented. Section 4 covers RF/microwave communication equipment interference problems and provides some interference control guidelines.

10.2 The Friendly Field of RF and Microwaves

According to information spread throughout the literature[1-5], microwaves and other types of RF frequencies so far have not been proved to have any long-term effects on human skin at exposure rates under 10 mW/cm^2 (the previous ANSI rate). On the other hand, various articles appearing in popular press often refer to research efforts in the Soviet Union and China, and refer especially to cases of "microwave sickness." However, these articles mention cases where workers had been exposed to unstated frequencies for unstated periods of time. There is, therefore, considerable disagreement in these instances. If applied under controlled conditions, microwave power not only is harmless, it also has therapeutic advantages over other techniques.

The U. S. Federal Communications Commission has authorized several frequency bands, including 13.56, 27.12, 40.68, 915 and 2,450 MHz, for *industrial, scientific* and *medical (ISM)* uses. However, the use of non-ISM frequencies demands extremely efficient shielding of the therapeutic area to avoid interference with communications facilities using nearby frequencies.[5]

10.2.1 The ANSI Safety Standard for Exposure to RF and Microwaves[2]

The growing interest in radiation monitors can be traced in part to the increasing awareness that electronic devices act as emitters of radiation. The investigative work is performed by such groups as the *American National Standards Institute (ANSI)*, the *Bureau of Radiological Health (BRH)* and the Department of Labor's *Occupational Safety and Health Administration (OSHA)*. In 1982, ANSI approved a new safety standard for exposure to RF and microwave energy fields covering frequencies from 300 kHz to 100 GHz (Fig. 10.1). ANSI's earlier standard had minimum levels of 10 mW/cm^2; the new standard goes down as far as 1 mW/cm^2 in the range of 30 to 300 MHz. These levels are the mean squared electric (E^2) and magnetic (H^2) field strengths and are given in terms of the equivalent plane-wave, free-space power density, measured at a distance of at least 5 cm from an emitter. The new

Exposure Safety

ANSI standard (C95.1–1982) does allow one exception to the 1 mW/cm^2 level: when the radio frequency input power of the radiating emitter is 7 W or less. When hazardous fields are composed of different frequency components, the ANSI standard allows that the fraction of the equivalent density level that is incurred within each frequency group should be determined, and the sum of these fractions should not exceed unity.

The ANSI standard has resulted in the development of a *Radio Frequency Protection Guide (RFPG)* which is based on the amount of electromagnetic (nonionizing) energy which the human body has been able to tolerate safely in experiments. This amount is referred to as the *specific absorption rate (SAR)*—the rate at which nonionizing energy is imparted to biological tissue.

The ANSI standard is based on a SAR of 0.4 W/kg as the upper limit, when averaged over the entire body. The spatial peak SAR is 8 W/kg when the average absorption is measured for one gram of tissue.

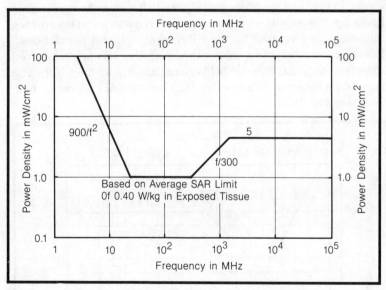

Figure 10.1—Updated ANSI Standard for RF and Microwave Radiation Exposure

10.2.2 New UK Limits for Exposure to RF and Microwaves[6]

New UK limits for exposure to RF and microwave radiation have been proposed in a consultative document from the National Radiological Protection Board. The publication proposes a mean specific energy absorption rate in the whole body of 0.4 W/kg for microwave and RF radiation. The current UK limit of 1 W/kg, recommended by the Home Office and Medical Research Council, has stood for around 20 years. Presumably the Health and Safety Executive will use the document in its final form as the basis for new regulations.

Handheld radio transmitters, intruder alarms and proximity devices emitting less than 7 W, "may be regarded as harmless," but they should be designed so that they cannot deliver more than 4 W/kg to the eye for long periods. RF and microwave hazards to people with pacemakers are unlikely, provided that the limits shown in Fig. 10.2 are observed. Figure 10.2 shows the permissible limits for continuous exposure to radio frequency and microwave radiations as proposed by the NRPB. For "general populations," levels are almost identical to those of the recently approved American National Standards Institute safety guidelines (C9). The curve dips between 30 and 3,000 MHz because of body resonance (see also Fig. 10.1).

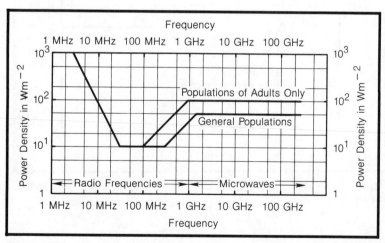

Figure 10.2—Diagram of RF and Microwave Limits Proposed by NRPB (UK)

10.2.3 Therapeutic Effects and Shielding Requirements

As previously mentioned, radio-frequency radiation applied under control conditions has therapeutic advantages over other techniques. Radio-frequency radiation in medicine—widely used to help heal bone fractures—is also showing promise for healing flesh wounds, regenerating nerves, and treating advanced cancer.[5]

Living tissues appear as lossy dielectrics to electromagnetic waves. Figure 10.3 shows the penetration depth of microwave power in human tissue with frequency[4] for muscles, bones, etc. It can be seen that microwave radiation can penetrate the body's fat layers and reach bones and other areas a few centimeters inside of muscles.[3,7]

The use of non-ISM frequencies demands proper and effective shielding of the therapeutic area to avoid interference with communications facilities using nearby frequencies. At the hyperthermic treatment center at the University of Indiana, for example, the entire treatment room is enclosed in grounded, close-mesh wire screening. Nine 200 W generators feed into nine folded dipole antennas arranged in three concentric rings around the torso of the patient. The power level is adjustable, but typically about 1 kW of RF output power is used. This amounts to about 13 W/kg or 100 mW/cm^2 for the period during which the source is operating.[5]

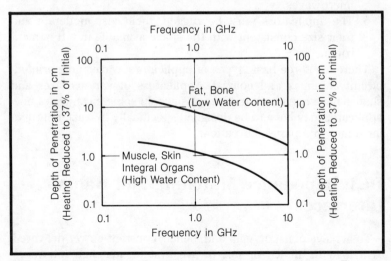

Figure 10.3—Penetration Depth of Microwave Power in Human Tissues with Frequency

10.2.4 RF/Microwave Applicators[3,8]

In most cases, the RF and microwave applicators or antennas that direct the radiation into the body must be custom made. When considering applicators for producing localized hyperthermia with such waves, several considerations must be kept in mind:
1. The applicators must be able to handle the RF/microwave power required to raise the temperature of the tumor or tumors to be treated to the effective hyperthermic range.
2. The design of the applicators must minimize the amount of power delivered to healthy tissues. With applicators operated at 915 or 2,450 MHz, the energy can be concentrated into well-defined volumes of tissues, but the depth of penetration into muscle tissues at these frequencies is small. Conversely, with applicators operating at the three lowest frequencies, the energy can penetrate deeply, but the wavelength in tissues at these frequencies is so large that diffraction effects lead to substantial spreading of the energy.
3. The applicator design must be consistent with the physical comfort of a patient who has to undergo one or more treatments at a time, each treatment lasting up to one hour.
4. Radiation into free space from the applicators must be kept to a minimum to protect the patient and the technician administering the treatment from unnecessary exposure to microwaves.
5. The applicators must be rugged, their cost moderate and their size consistent with the space available in a treatment room.

There are three basic types of applicators: waveguide, printed-circuit antennas and coaxial. Applicators of various sizes and shapes have been built in these designs. In some instances, a new applicator may have to be designed specifically to treat a particular tumor in a particular patient.

10.3 Impedance Matching and Balun Interfaces

To achieve constant gain and, hence, constant power delivered to a load in an RF or microwave system, matched interfaces between amplifiers are required. This is of particular importance at the interface between the amplifier and the antenna.

10.3.1 Effects of Mismatch on Amplifier Performance at High Frequencies[9]

A mismatched load not only will result in varying power delivered to the load, but it also will affect the amplifier delivering the power, modifying the current and voltage swings within the amplifier and also the dissipation in the active devices. Designing amplifiers to withstand the additional voltage swings due to a mismatch, without increasing distortion, reduces efficiency. Designing for increased dissipation reduces the output power capability for a given active device. These aspects are important in the power stages of an amplifier chain, where efficiency is more important. An amplifier having a matched output impedance reduces these effects.

10.3.2 Simple Matching Networks[10]

Figure 10.4 shows a two-reactance matching network. Matching or exact transformation from R_2 into R_1 occurs at a single frequency f_0. At f_0, X_1 and X_2 are equal to:

$$X_1 = \pm R_1 \sqrt{\frac{R_2}{R_1 - R_2}} = R_1 \frac{1}{\sqrt{n-1}} \quad (10.1a)$$

$$X_2 = \mp \sqrt{R_2(R_1 - R_2)} = R_1 \frac{\sqrt{n-1}}{n} \quad (10.1b)$$

At $f_0 : X_1 \times X_2 = R_1 \times R_2$ \quad (10.2)

X_1 and X_2 must be of opposite sign. The shunt reactance is in parallel with the larger resistance.

If X_1 is capacitive and consequently X_2 inductive, then:

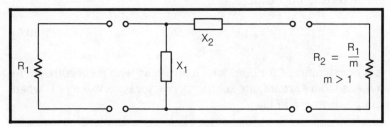

Figure 10.4—Two-Reactance Matching Network

$$X_1 = -\frac{f_0}{f} R_1 \sqrt{\frac{R_2}{R_1 - R_2}} = -\frac{f_0}{f} R_1 \frac{1}{\sqrt{n-1}} \quad (10.3a)$$

$$X_2 = \frac{f}{f_0} \sqrt{R_2(R_1 - R_2)} = \frac{f}{f_0} R_1 \frac{\sqrt{n-1}}{n} \quad (10.3b)$$

If X_1 is inductive and consequently X_2 capacitive, the only change required is a replacement of f by f_0 and vice versa. The L-section has low-pass form in the first case and high-pass form in the second case. The Q of the circuit at f_0 is equal to:

$$Q_0 = \frac{X_2}{R_2} = \frac{R_1}{X_1} = \sqrt{n-1} \quad (10.4)$$

For a given transformation ratio n, there is only one possible value of Q. On the other hand, there are two symmetrical solutions for the network which can be either a low-pass filter or a high-pass filter. The frequency f_0 does not need to be the center frequency, $(f_1 + f_2)/2$, of the desired band limited by f_1 and f_2. In fact, it could prove interesting to shift f_0 toward the high band edge frequency f_2 to obtain a larger bandwidth W, where:

$$W = \frac{2(f_1 + f_2)}{f_2 - f_1} \quad (10.5)$$

This will, however, be at the expense of poorer harmonic rejection.

In the above analysis, the inductance was expected to be realized by a lumped element. A transmission line can be used instead, as shown in Fig. 10.5. As can be seen from the computed selectivity curves for the two configurations in Fig. 10.5, transmission lines result in a larger bandwidth. The gain is important for a transmission line which has a length $L = \lambda/4$ ($\theta = 90°$) and a characteristic impedance:

$$Z_0 = \sqrt{R_1 \times R_2} \quad (10.6)$$

It is not significant for lines which are short with respect to $\lambda/4$. In this case there are infinite solutions, one for each value of C when using transmission lines.

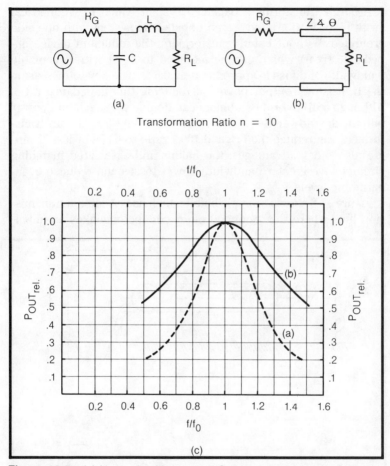

Figure 10.5—(a) Network with Lumped Constraints, (b) Network with a Transmission Line and (c) Bandwidth of L-Section for Both Cases with Transformation Ratio n = 10

10.3.3 Balun Transformers[11]

Using single RF power transistors, it is not possible to satisfy modern equipment's design criteria. Therefore, several devices in separate packages, or in the same package (balanced, push-pull or dual transistors), must be coupled to obtain the required amplifier output power. Since high-power transistors have very low impedance, designers are challenged to match combined devices to a load.

RF/Microwave Interfaces

A *balun* transforms a balanced system which is symmetrical (with respect to ground) to an unbalanced system with one side grounded. Without balun transformers, the minimum device impedance (real) which can be matched to 50 Ω with acceptable bandwidth and loss is approximately 0.5 Ω. The key to increasing the transistors' output power is reducing this impedance ratio. Although 3 dB hybrid combiners can double the maximum power output, they lower the matching ratio to only 50:1. Balun transformers can reduce the original 100:1 ratio to 6.25:1 or less. There are also other advantages: the baluns and associated matching circuits have greater bandwidth, lower losses and reduced even-harmonic levels.

Figure 10.6 shows a simple balun which uses a coaxial transmission line, where the grounded outer conductor makes an unbal-

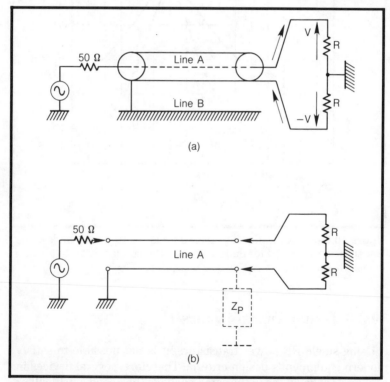

Figure 10.6—Simple Balun (a) Coaxial Design and (b) Equivalent Circuit (Line A: Z_A, Θ_A; Line B: Z_B, Θ_B; $Z_P = jZ_B \tan\Theta_B$, $Z_P \gg R$)

anced termination, and the floating end makes a balanced transmission. Charge conservation requires that the currents on the center and the outer conductors maintain equal magnitudes and a 180° phase relationship at any point along the line. By properly choosing the length and the characteristic impedance, this balun can be designed to match devices to their loads. In the case shown, if $\delta_A = 90°$, the matching condition is:

$$Z_A^2 = (2)(R)(50) = 100R \tag{10.7}$$

The basic balun can be made perfectly symmetrical by adding a second coaxial line (Fig. 10.7a). In this symmetrical coaxial balun

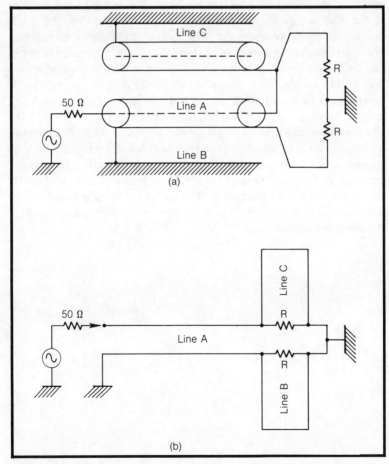

Figure 10.7—Symmetrical Coaxial Balun: (a) Coaxial Design and (b) Equivalent Circuit (Line A: Z_A, Θ_A; Line B: Z_B, Θ_B; Line C: Z_C, Θ_C)

the bandwidth, in terms of the input *voltage standing wave ratio (VSWR)*, limited by the transformation ratio 50/2R and the leakages which are represented by lines B and C. If $Z_A = 50\ \Omega$ and $R = 25\ \Omega$, the bandwidth is constrained only by the leakage. The equivalent circuit for the symmetrical balun (Fig. 10.7b) shows the effect of the leakages (lines B and C) on its performance. A broadband balun can be obtained by using a relatively high characteristic impedance for these leakage lines. In theory, the construction of the baluns ensures perfect balance.

Another balun design adds two identical coax lines to the previous design, resulting in the network shown in Fig. 10.8a. The inputs of the identical lines are connected in series to the output of the first balun. By putting their outputs in parallel, the final output becomes symmetrical. The output impedance is halved. The equivalent circuit of this design (Fig. 10.8b) indicates that its bandwidth, in terms of input VSWR, is limited by the transformation ratios of the first and second sections and the leakages represented by lines B, C, E and G. If the balun is designed with $Z_A = 50\ \Omega$, and $Z_D = Z_F = 25\ \Omega$, and if the load 2R is set at $2 \times 6.25\ \Omega$, all of the transmission lines will be connected to their characteristic impedances. In this case, the bandwidth will be limited by the leakage alone, and a broadband balun can be obtained by choosing lines B, C, E and G with relatively high impedance, and a $\lambda/4$ length for the center frequency. The balun achieves a transformation from $50\ \Omega$ to twice $6.25\ \Omega$ without causing a standing wave in the coaxial cables.

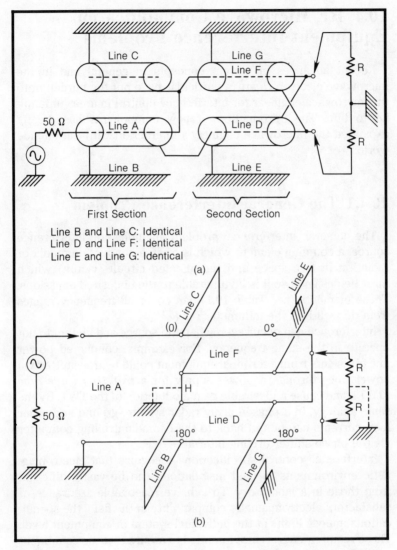

Figure 10.8—(a) Two-Section Coaxial Balun and (b) Equivalent Circuit

10.4 RF/Microwave Communication Equipment Interference Problems

Recent advances in system concepts in general and in the microwave electronics art in particular have put the burden onto the microwave engineer for interference control in areas unfamiliar to him. The introduction of new components has greatly expanded the microwave engineer's interaction with the outside world.

10.4.1 The General Interference Problem

The general interference problem includes an interference source, a coupling element which is often a conduction media or radiation in free space and the affected circuit (victim) which must be designed so it is invulnerable to the undesired emissions. These elements (see Table 10.1) can cover all frequency ranges from dc well into the millimeter range.[12]

Interference can occur even when the source and victim do not operate in the same frequency. For example, conducted power line transients from a terminal equipment could be transmitted via power lines through a power supply for a *traveling wave tube (TWT)*, and cause undesirable gain modulation of the TWT. By the same analogy, in a phased array radar system, ground and other loop currents can be fed back to the beam-controlling computer via the phase shifter driving interfaces.[13-15]

Interference control specifications recognize that electromagnetic environments in field installations can be vastly different from those in a laboratory. To achieve reasonable assurance of satisfactory electromagnetic compatibility in the field, the specifications impose limits at the individual system or equipment level. These requirements are not intended to assure that the equipment will operate in its own electromagnetic environment (that is handled by the equipment's performance specification), but that it will operate in proximity to other equipment similarly controlled, with minimal possibilities for interference. Interference control specifications basically call for two types of control. One is suppression of the sources of interference themselves. The other requires that the equipment be invulnerable, or nonsusceptible, to specified levels of ambient interference.[12]

Table 10.1—Basic Elements of the RF/Microwave Interference Problem *(Adapted from Ref. 12)*

Interference Sources	**Natural Sources** • Atmospheric Static • Lightning • Galactic **Man-Made Sources** • Radar and Communications Transmitter, Desired and Undesired Emissions • Electrical Machinery, Radiated and Conducted Emissions • Neon, Fluorescent Lights • Switches, Solenoids • Electronic Equipment
Coupling Element	• Antennas • Mutual Inductance • Mutual Capacitance • Common Power Supply Impedance
Affected Circuit (Victim)	• Radar Communications Receivers • Computers • Terminals, Indicators • Measuring Instruments • Electro-Explosive Devices

10.4.2 Near-Field Communication[16]

To provide *unlicensed* operation for devices like garage door openers, very low power operation is required by the FCC. This is where *near-field* operation comes in. It may be desirable, at this point, to define *near-field*. An RF current in a transmitted loop antenna produces an oscillatory magnetic field. This field, in turn, induces a proportional current in a receiver loop antenna placed in it. Thus, the transmitter loop-receiver loop system can be thought of as a very loosely coupled (*leaky*) transformer. In this case, the field varies as the inverse cube of the antenna separation for distances considerably less than a wavelength. At distances over one wavelength, the usual inverse low applies. Thus, using very low frequencies (long wavelengths), a usable signal level can be obtained while still satisfying the FCC requirements. Table 10.2 shows the maximum allowable field strength for nonlicensed operation.

There are a host of applications where the *near-field* approach provides important advantages. Basically, they include low-power, short-range, systems which can be divided into two main categories: (1) systems which cannot exceed a certain field strength at a specified distance and (2) systems where partial security of transmission is desired.

Table 10.2—Maximum Allowable Field Strength for Nonlicensed Operation (FCC Regulations-Part 15)

Frequency-kHz	Distance-meters	Field Strength
10-490	300	2400/frequency
510-1600	30	24,000/frequency
Frequency-MHz		
70-130	30	50
130-174	30	50-150 (linear interpolation)
174	30	150
260-470	30	150-500 (linear interpolation)
470 and above	30	500

10.4.3 Radiating RF Cables

Radiating RF cables are transmission lines whose outer conductor carries a series of equally spaced circular apertures. These apertures allow an RF signal to leak in or out of the transmission line. Therefore, radiating cables can be considered as long antennas for use in confined places or, alternatively, in open areas.

Using radiating RF cables, the useful frequency range extends to 500 MHz, although the most efficient operation occurs at frequencies between 40 MHz and 200 MHz. Applications include: radio communication systems for motorway and train tunnels, mines, underground railways, staff or personnel paging systems in large buildings and passageways and monitoring of trains on line-sections and at stations.[17]

According to Ref. 17, in experiments with lossy depositions over the outer cable surface (e.g., salt, water, oil, metallic oxides, carbon and other substances of high dielectric constant and high loss tangents), applied in the form of water based coatings, the attenuation did not increase more than 0.2 dB/100 m at 160 MHz. The corresponding coupling loss was −3.5 dB. These cables are therefore not particularly sensitive to a normal environment. However, periodic cleaning of the cables is recommended, since excessive accumulations will have an increasingly detrimental effect on the system attenuation. The normally used tunnel cleaning materials and methods have no electrical or material influence on the cable.

10.4.4 Microwave Transmission System for LANs

A radio-based system which provides the digital communication links between multiple subscriber premises and a node of a large telecommunications network is referred to by the FCC as a *digital termination system (DTS)*. In the FCC's August 1979 *Notice of Proposed Rulemaking and Inquiry*, the commission recognized the public need for a new high-bandwidth digital local distribution system for use by competing regional and national networks. It proposed allocations of frequency bands 10.57 to 10.6 GHz and 10.635 to 10.665 GHz to DTS networks (a total of 60 MHz) with 30 MHz to be held in reserve bands of 10.6 to 10.615 GHz and 10.665 to 10.68 GHz. Although some changes followed thereafter for the final allocation, it was the first step in developing DTS intended to interconnect many geographically separated digital information devices.

There are two potential types of transmission architectures where at each subscriber location users are connected to a *network interface unit (NIU)*.[18] In the first architecture type, all pairs of NIUs are connected directly by *point-to-point* microwave links. In the second architecture type, each NIU is connected to a central location called a *local node*. This centralized architecture can use dedicated point-to-point uplinks and downlinks, or it can use multiple access point-to-point uplinks and a broadcast downlink (Fig. 10.9)

In the case of RF/microwave communication, fading will occur when the desired signal arrives at the receiver by more than one propagation path. In the office environment, multiple transmission paths may be due to reflections from walls, ceiling, objects, etc. At some frequencies, the relative time delay between these paths may be such that the signal components may cancel one another at the receiver while reinforcing one another at other frequencies. These will result in selective *fading* and *intersymbol* interference.

RF/Microwave Interfaces

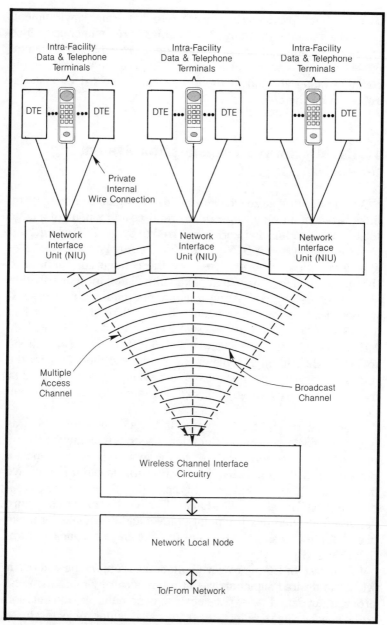

Figure 10.9—Microwave Transmission System for LANs

In general, when using a microwave transmission system in a confined place like an office environment, interference and noise are not easily predictable and avoidable because of multiple signal reflections. Other factors include the change of positions of handheld wireless terminals or telephone sets (*telsets*) and the switching of various devices, thereby creating transient phenomena.[19] *Spread-spectrum* techniques which require no knowledge of interference and noise conditions can be used to suppress noise and interference of various forms which may have the same frequency, same polarization and same direction of arrival as that of the desired signal.

10.4.5 Aerospace EMC: Tests Specifications Comparison[20]

The high density of electronic systems operating at RF and microwave frequencies, as well as the complex cabling and the reliability requirements, make aircraft and spacecraft difficult EMC environments.

Specifications are available which provide procedures for EMC testing of aerospace electronics systems and subsystems. The tests are aimed both at measuring the interference generated by the systems and determining their sensitivity to interference. The specifications provide test procedures and limits for *conducted emission (CE)*, *conducted susceptibility (CS)*, *radiated emission (RE)* and *radiated susceptibility (RS)*. While in military applications the term *requirements* is used, the civilian specification documents refer to *guidance* (Table 10.3). Frequently, however, government aviation authorities require civilian equipment to comply with the applicable specification from *Aeronautical Radio, Inc. (ARINC)* or the *Radio Technical Commission on Aeronautics (RTCA)*.

There are many similarities between the RTCA and MIL-STD documents, including procedures, test setups and types of transducers used during tests. Table 10.4 shows a basic comparison of MIL-STD and RTCA tests. For most tests, the limit values in RTCA DO-160 are slightly less strict than those in MIL-STD-461. The RTCA standard is more stringent, however, for broadband interference generated by switch circuitry which is simulated by so-called spike generators. The transducers required for EMC testing

RF/Microwave Interfaces

depend on the type of test being done. For conduction tests, a transformer, current probe or line-impedance stabilization network is applicable. For radiation tests, loop antennas may be used for magnetic field measurements at lower frequencies. At higher frequencies, electric field measurements can be made with rod antennas, long wire antennas in a shielded room, biconical antennas, conical log spirals or horn antennas.

Some EMI tests can be performed outdoors, in the natural environment of the equipment. Most EMI testing, however, is best performed within a shielded anechoic chamber.[21] Regardless of the specification used, the purpose of EMC testing is to determine if the equipment under test generates a level of interference higher than the susceptibility threshold of other equipment. In practice, if the equipment meets the specifications, EMC is ensured. The specifications are severe, however, and are often hard to meet, so waivers sometimes are used.

Table 10.3—The Major Aerospace EMI Specifications*

Military ("Requirements")		Civil Industry ("Guidance")	
Aircraft Power (Incl. Transients)	EMC	Aircraft Power (Incl. Transients)	EMC
MIL-STD 704B (1975)	MIL-E-6051D (Sep. 1967)	ARINC 413A (Sep. 1976)	RTCA DO-160B (July 1984)
		Compatible with MIL-STD 704B and RTCA DO-160B	(Environmental Spec)
For EMC Specs Refers to			
MIL-STD 461 MIL-STD 462	EMC Procedures and Limits	RTCA DO-160B Sects. 17-21	

*Reproduced from Microwaves & RF by Permission; Copyright © 1985, Hayden Publishing Co., Inc.

Table 10.4—Basic Comparison of MIL-STD and RTCA Tests*

MIL-STD		RTCA	
Conducted Emission			
CE01	30 Hz to 15 kHz Current: 130 dB(μA) max.	Para.	
CE03	15 kHz to 50 MHz: 86 to 20 dB(μA) NB** 130 to 50 dB(μA/MHz) BB	21.3	15 kHz to 30 MHz: 75 to 20 dB(μA) NB 120 to 50 dB(μA/MHz) BB
CE07	Spikes: AC: ±50 percent V_{nom} DC: +50/−150 percent V_{nom}		
Radiated Emission			
RE01	30 Hz to 50 kHz H-field: 140 to 20 dBpT		
RE02	14 kHz to 10 GHz E-field 30 to 70 dB(μV/m) NB 110 to 65 dB(μV/m-MHz) BB	21.4	15 kHz to 1215 MHz E-field: 35 to 62 dB(μV/m) NB 98 to 68 dB(μV/m-MHz)
Conducted Susceptibility			
CS01	30 Hz to 50 kHz: 5 to 1 V, CW	18	10 Hz to 150 kHz DC: 1.4 V_{max} AC: 5 percent V_{nom}
CS02	50 kHz to 400 MHz: 1 V, modulated	20.4	90 kHz to 30 MHz: 0.5 V_{max}, modulated
CS06	Spikes on power leads: 200 V, 10 μs, 0.15 μs	17	Spikes: 600 V, 10 μs, t_r < 2 μs
CS09	60 Hz to 100 kHz: Structure current, < 1 A		
Radiated Susceptibility			
RS01	30 Hz to 50 kHz H-field 120 to 76 dBpT	19.3.2	400 Hz to 15 kHz H-field induction: 30.0 to 0.8 A-m
		19.3.3	E-field induction: 400 Hz, 1800 V-m
RS02	Spikes, induction: 20 A, 10 μs	19.3.4	Spikes, induction: 600 V, 0.2 to 10.0 μs
RS02	Power frequency, induction field: 20 A	19.3.1	Power frequency induction: 20 A
RS03	14 kHz to 10/40 GHz E-field 5 to 10 V/m	20.5	H-field, 15 kHz to 35 MHz E-field, 35 to 1215 MHz: 0.2 to 1 V/m
**NB = narrowband interference; BB = broadband interference			

*Reproduced from Microwaves & RF by Permission; Copyright © 1985, Hayden Publishing Co., Inc.

10.4.6 Some Guidelines for Avoiding Interference Problems in Microwave Designs

Below, some interference control guidelines are given which will aid microwave systems in surviving in real-world electromagnetic environments:[12]

1. *Apply interference control at the local level.* Powerful sources such as comb generators and local oscillators, and sensitive solid state circuitry such as amplifiers, should be individually packaged in their own shielded modules. All interference lines should be shielded or filtered to preserve the shielding integrity of the enclosure.

2. *Avoid substituting shields for filters.* Secondary dc power, bias and control lines should be filtered at the interface to the shielded enclosure with suitable feedthrough filters. If shielded lines are used where filters could be used equally well, this will result in unnecessary extension of the shielded enclosure.

3. *Use adequate metal thickness for shielded enclosures.* The intrinsic shielding ability of the basic metal is rarely a consideration at microwave frequencies. Above a few MHz, the theoretical shielding effectiveness of .762 mm thick aluminum against radiated plane waves is in excess of 200 dB. However, actual enclosures do not provide such high protection due to leakage, particularly through seams. Covers and enclosures should be sufficiently rigid at their interface to prevent waviness at the contacting surface. The use of anodizing as a protective finish on mating surfaces requiring electrical continuity must be avoided since anodize is an insulator.

4. *Use conductive gasketing.* For removable covers, conductive metal gaskets or spring finger contacts must be used. A suitable gasket or spring finger contact, when correctly applied, can compensate for surface unevenness and assures a nearly continuous electrical bonding between mating surfaces.

5. *Isolate integrated microwave assemblies.* Isolation is a problem in integrated assemblies employing a number of microwave and associated solid state components in a single enclosure. It is possible that slight asymmetry in the microwave portion of the system will create higher order modes which can then propagate throughout the structure with very

low loss, especially where the structure becomes resonant. Conducting posts, which electrically tie the ground planes together in proximity to potential sources, effectively suppress such higher order modes. Microwave absorbers, strategically installed within the enclosure, will effectively suppress resonances while not affecting intended transmissions.
6. *Look for parasitic oscillations and spurious responses in receivers.* Various switching elements and amplifiers can oscillate with harmonics going up to microwave frequencies. Besides interference problems, such parasitics usually present reliability problems. Sometimes these phenomena cannot be observed using an oscilloscope; therefore, a special probe and a spectrum analyzer should be used to scan the circuit in order to see if parasitics exist. On the other hand, detailed spurious response analyses are necessary for the receivers in the early design phase. If spurious problems are revealed early enough, they can be eliminated by a correct choice of IF frequency amplifier-type and component screening.
7. *Eliminate cable crosstalk.* In an actual installation like a large phased array radar system, interface lines (and there are several thousands of these) can pick up substantial levels of power line interference. Such signals could cause undesirable effects to varactors and PIN microwave diodes. They are best eliminated by routing the signal on shielded twisted pairs and using opto-isolators.

10.5 References

1. M. Smith, *Microwave Health Arguments Intensify*, MSN, May 1981, p. 19.
2. J. Browne, *Radiation Monitors Measure Potential Health Hazards, Microwaves and RF*, March 1983, p. 121.
3. F. Sterzer, et al., *Microwave Apparatus for the Treatment of Cancer, Microwave Journal*, January 1980, p. 39.
4. C. Gupta, *Microwave in Medicine, Electronics and Power*, IEE, May 1981, p. 403.
5. R. E. Shupe and N. B. Horback, *The Friendly Fields of RF, IEEE Spectrum*, June 1985, p. 64.
6. News Section of Wireless World, *Proposals for Non-Ionizing Radiation Limits, Wireless World*, April 1983, p. 56.
7. C. C. Johnson, and A. W. Guy, *Non-Ionizing Electromagnetic Wave Effects in Biological Materials and Systems, Proceedings IEEE*, Vol. 60, June 1972, pp. 692–718.
8. F. Sterzer, et al., *RF Therapy for Malignancy, IEEE Spectrum*, December 1980, pp. 32–37.
9. R. E. Gerard, *Mismatch Effects in hf Wideband Systems*, Communications Engineering International, April 1980, p. 10.
10. B. Becciolini, *Impedance Matching Networks Applied to R-F Power Transistors, Application Note AN-721*, Motorola Semiconductor Products, Inc.
11. R. Basset, *Three Balun Designs for Push-Pull Amplifiers, Microwaves*, July 1980, pp. 47–52.
12. R. J. Mohr, *Controlling Interference in Microwave Design, Microwaves*, November 1971, p. 31.
13. C. J. Georgopoulos, *Interaction of High Power PIN Diodes and Driving Circuitry During Forward-Bias Switching, IEEE Journal of Solid-State Circuits*, Vol. SC-11, No. 2, April 1976, pp. 295–302.
14. C. J. Georgopoulos, *Design of PIN Diode Drivers for Phased Array Radars: Some New Approaches*, Proceedings of 1983 International Radar Symposium, Bangalore, India, October 10–13, 1983, pp. 144–148.
15. C. R. Boyd, *A Dual-Mode Latching Reciprocal Ferrite Phase Shifter, IEEE Transactions on Microwave Theory and Techniques*, Vol. MTT-18, No. 12, December 1970, pp. 1119–1124.
16. A. B. Przedpelski, *"Near Field" Communication, RF Design*, March/April 1984, p. 50A.

17. N. Forbes-Marsden, *Radiation RF Cables in Communication Systems, Electronic Engineering*, February 1980, p. 73.
18. L. M. Niremberg, and C. T. Wolverton, *Microwave Transmission Systems for Local Data Networks, Telecommunications*, June 1980, p. 47.
19. C. J. Georgopoulos, *Spread-Spectrum Techniques for Reliable and Private Communications in the Office of the Future*, Proceedings of the DIGITECH '84 IMACS European Meeting, Patras, Greece, July 9–12, 1984.
20. O. B. M. Pietersen, *The Right Combination Unlocks Aerospace EMC, Microwave & RF*, October 1985, pp. 73–78.
21. J. B. Schultz, *Radio Frequency Chamber Improves LAMPS MK III Testing, Defense Electronics*, November 1985, p. 53.

General References

- J. Toler and V. Popovic, *Methods and Approaches for Studying Biological Effects of Radio Frequency Radiation: An Overview*, EMC—5th Symposium and Technical Exhibition on: Electromagnetic Compatibility, Zurich, Switzerland, March 8–10, 1983.
- N. H. Steneck, et al., *The Origins of U.S. Safety Standards for Microwave Radiation, Science*, Vol. 208, June 13, 1980.
- C. Y. Ho, *New Analysis Technique Builds Better Baluns, Microwaves & RF*, August 1985, pp. 99–102.
- D. Scherer, et al., *Low Noise RF Signal Generator Design, Hewlett-Packard Journal*, February 1981, pp. 12–22.
- R. E. Easson, *Making Difficult Measurements of Noise Power Ratios on Transmitters, MSN*, February 1981, p. 88.
- U. S. Department of Commerce/National Bureau of Standards, *An Electric and Magnetic Field Sensor for Simultaneous Electromagnetic Near-Field Measurements-Theory*, NBS Technical Note 1062, April 1983.
- H. W. Heather, *Application of Nomographs for Analysis and Prediction of Receiver Spurious Response EMI*, Naval Air Test Center, Patuxent River, MD, July 23, 1985.
- R. E. Burns, *A High-Performance Signal Generator for RF Communications Testing, Hewlett-Packard Journal*, pp. 4–6, December 1985.
- B. Prusad and B. N. S. Babu, *EMP Effects on the Performance of Direct Sequence Spread-Spectrum Communication System, IEEE Transactions on Communications*, Vol. COM-32, No. 12, December 1984, pp. 1251–1259.

- C. J. Georgopoulos and V. Makios, *PIN Diode Reverse-Bias Switching Via Inductive Discharge, Electronics Letters*, Vol. 14, No. 23, November 9, 1978, pp. 723–725.
- E. Freeman, *Interference Suppression Techniques for Microwave Antennas and Transmitters*, Artech House, Inc., Dedham, MA, 1982.
- G. Bostick, *Damping Spurious Microwave Response With Absorbing Materials, EMC Technology*, April–June 1985, p. 21.

Index

A

Acquisition time 178
Active/passive termination 164
Active pull-down. 164
Advanced Low-power Schottky (ALS) 29
Advanced Schottky (AS) 29
Alarm Interfaces 290
Alarm Management. 292
Ambient level. 227
Anechoic Chamber 21
Antenna farm. 7, 282, 283
Aperture time 178
Attenuation 73
Automation Interfaces 207

B

Balun interfaces 304
Balun Transformers 307
Bridge amplifier. 208
Broadband ratio noise 228
Built-in testability 215
Bus
   acceptor handshake function (AH) 144
   amplifier 147
   Bitbus 149, 158

Index

control 141
controller function (C)................................. 145
data... 141
device trigger function (DT)............................ 145
extender.. 147
Futurebus... 159
GPIB.. 143
interface... 139
line... 117, 166
listener function (LE).................................. 144
MIL-STD-1553B 158
Nubus... 158
organized systems....................................... 163
parallel poll function (PP)............................. 145
power 141
Q-bus... 159
remote/local function (RL) 145
service request function (SR)........................... 145
S-100 158
source handshake function (SH) 143
STD... 140
STE bus 159
talker function (T or TE)............................... 144
VME bus... 152
VERSAbus 158

C

Cable-shield qualities 290
Capacitive filter....................................... 89
CCITT V24 4
Characteristic impedance 49, 80
Cheapernet... 267, 287
Chopper stabilizer amplifier 210
CISPR
    limits.. 19
    recommendations..................................... 18
CMOS
    ESD protection 56
    family 33
    interconnections 51
    latch-up.. 51

Index

transfer characteristics 51
CMOS/ECL interface. 64
CMOS/TTL interface. 62
Coax cable 82, 83
Common-mode rejection 88, 211
Common-mode rejection rate (CMRR) 196, 197, 211
Computer grade connectors 128
Conducted
  emission (CE) 317
  emission tests 228
  susceptibility (CS). 317
Conduction loss 73
Connector interface 89
Coupling path 10
Crosstalk 73, 88

D

Data
  acquisition system (DAS) 173
  circuit equipment (DCE) 4, 104
  relaying. 201
  terminal equipment (DTE) 3, 4, 104
  terminal relay 4
Difference amplifier 208
Differential backplane transceivers 166, 167
Differential input multiplexers 176
Digital multiplexed interface (DMI). 109, 110
Digital signal line model 73
Digital-to-analog (D/A) converters 181
Drivers
  differential 90
  single-ended 90

E

ECL family 31, 32, 40
ECL-to-ECL connection 48
ECL/TTL interfaces. 60
Effective dielectric constant 49
EIA
  RS-232, -232-C 4, 71, 93, 96, 97, 111, 112, 196

327

Index

    RS-422, -422-A 71, 94, 95, 97, 112
    RS-423 .... 71, 96, 112
    RS-485 .... 71, 94, 97
    RS-449-A 111
Electromagnetic
    compatibility (EMC) 1, 8, 11
    interference (EMI) 6, 9, 55
    pulse (EMP) 10, 132
    radiation (EMR) 11
    receptors 10
    sources 10
Electrical overstress (EOS) 13
Electrometer amplifier 210
Electrostatic discharge 13
Emission 227
Emitter-Coupled Logic (ECL) 31
ESD damage thresholds 41
Ethernet 267, 287
Expansion system bus (iSBX) 149
External signal conditioners 150

F

FCC
    limits 15, 16
    rules 226
    standard of measurements 227
Feedback capacitor 89
Fiber optic(s)
    bandwidth budgeting 257
    cladding 243
    connectors 247
    core 243
    first window 244
    flux budgeting 257
    index profile 243
    link 241
    material dispersion 244
    model dispersion 244
    radiation effects 249
    second window 244
    sensors 245

Index

    snap-in link 255
    third window 244
Filter
    capacitor 89
    feedthrough 89
    interference 89
Floppy disk 122
Frequency division multiplexing (FDM) 271
Friendly fields 299

G

GaAs ICs 37, 41
GaAs radiation hardness 42
Gas discharge laser system 240
Ground plane 228

H

Half-duplex 108
Hazardous areas 202
High-speed logic 35
High-threshold logic (HTL) 45
Hi-Z state 28
HSCMOS/CMOS interface 65
Hybrid nuclear event detector (HNED) 133
Hysteresis 98

I

Immunity testing 20
Impedance matching 304
Inductive kick 129
Infrared 131, 235, 258
Instrument amplifier 188
Isolation amplifier 210, 213
Interactive operation 113
Interface(s)
    alarm 290
    automation 207
    balanced 4
    CCITT standards 3

Index

CMOS/ECL 64
CMOS/TTL 62
definition 2
device 2
ECL/TTL 60
EIA standards 3
HSCMOS/CMOS 65
intelligent peripheral 123, 127
man/machine 113
measurements 207
micro-to-mini 124
microwaves 302
peripherals 103, 104, 119
physical 3
specification 3
terminals 103, 104, 112
unbalanced 4
Interference filters 89

J

Johnson noise 9

K

Keyboard 113

L

LAN topology 268, 269
Laser power supply (LPS) 240
Light sources
    LEDs 236, 251
    injection lasers 251
    spectral matching 250
Line
    driver 90
    impedance stabilization network (LISN) 228
    receiver 90
    termination 76, 78
Local area network (LAN) 265
Local bus exchange 149

Local bus extension (iLBX) 149
Logic analyzers C 221
Logic families 25
Lossy ferrites 89
Low-power Schottky (LS) 29

M

Magnetic disk storage 119
Magnetic tape storage 119
Man/machine interface 113
Matching networks 305
Measurement interfaces 207
Microcomputer-Operated Process Control System . 186
Microstrip line 49, 76
Microwave exposure limits 301, 302
Modem 104, 109
Monitoring 290
Multibus I 149
Multibus II 149
Multibus system 148
Multicomputing 148
Multiprocessing 148

N

Narrowband signals 10
Narrowband radio noise 228
Near field 313
Noise
  atmospheric 9
  flicker 9
  Johnson 9
  man-made 9
  margin 27
  microphonic 9
  RF 9
  shot 9
  white 9

O

Open-site measurements 20
Operational amplifier 208
Optical isolators 235, 236
Optocouplers 131
Overshoot 48

P

Parallel system bus (iPSB) 149
Peripheral equipment 119
Peripheral interfaces 103, 104, 119
Photodetectors 253
PN junction 113
Propagation delay 73

Q

Quadrax cable 83, 84

R

Radiated
    emission (RE) 317
    emission tests 230
    radio noise 228
    susceptibility (RS) 317
Radiative EMI 11
Radiating RF cable 314
Radio-frequency interference (RFI) 9
Radio noise 228
Random noise 228
Reflection coefficient 73
Repeaters 97, 98
Resistive loss 73
Restoration 290
Reverberating chamber 20
RF exposure limits 301, 302
Ribbon cable 81
Rigid disk 122
Ringing 48

S

Sampling rate.. 178
Schottky clamped...................................... 29
Schottky TTL.. 28
Sensors... 187
Serial system bus (iSSB).............................. 149
Settling time... 178
Shielded cables....................................... 84
Shield termination 130
Signal conditioning................................... 150
Signal conditioning manifolds......................... 150
Signal sources.. 187
Single event upset (SEU).............................. 41
Skin effect 73, 85
Solenoid dropout 129
Spectrum analyzer 222
Spread-spectrum....................................... 317
Standards
    CISPR... 18
    Class A equipment 16
    Class B equipment 16
    EIA 3
    FCC... 15
    limits................................... 16, 17, 19
    VDE... 16
Storage module drive (SMD) 123
Streamer buffered interface 124
Stripline 49, 76
Susceptibility testing 20

T

TEM cells... 20
Temperature effects................................... 202
Terminal interfaces................................... 112
Test fixtures... 216
Thermocouples... 188
Toggle frequency...................................... 29
Token passing... 270
Transceivers.......................................97, 98
Transducers... 187

Index

Transducer amplifier 208
Transient radiation upset... 41
Transient suppression 129, 130
Transimpedance circuit... 254
Transmission media
    coaxial cable 272, 273
    fiber optic cable 272, 274
    flat cable... 80
    infrared (IR) carrier 272, 276
    radio-frequency (RF) carrier............................ 272, 274
    twisted pair cable...................... 49, 81, 272, 273
Triax cable.. 82, 84
Tri-state logic... 28
TTL family 26, 30, 40, 59
Twinax cable.. 82, 84
Twisted pair line.. 49, 81

U

Ultra-isolation transformer 284
Unused inputs 51

V

VDE limits 17
Virtual ground system 199
VME bus.. 152
Voltage isolation 193
Voltage-to-frequency (V/F) converters...................... 190

W

Winchester disk drive 123
Wire center 293
Wiring concentrators 293
Wire over a ground.................................... 258, 272
Wireless links 258, 272